GEOLOGY
UNDERFOOT
IN *Yosemite*
NATIONAL PARK

Allen F. Glazner and Greg M. Stock

2010
Mountain Press Publishing Company
Missoula, Montana

*The Geology Underfoot series presents geology with a
hands-on, get-out-of-your-car approach. A formal background in geology is not required for enjoyment.*

is a registered trademark of
Mountain Press Publishing Company.

Library of Congress Cataloging-in-Publication Data

Glazner, Allen F.
    Geology underfoot in Yosemite National Park / Allen F. Glazner and
Greg M. Stock.
        p. cm.
    Includes bibliographical references and index.
    ISBN 978-0-87842-568-6 (pbk. : alk. paper)
    1. Geology—California—Yosemite National Park. 2. Natural history—
California—Yosemite National Park. I. Stock, Greg M., 1973– II. Title.
    QE90.Y6G53 2010
    557.94'47—dc22
                        2010007356

PRINTED IN HONG KONG BY MANTEC PRODUCTION COMPANY

P.O. Box 2399 • Missoula, MT 59806 • 406-728-1900
800-234-5308 • info@mtnpress.com
www.mountain-press.com

To all of my geology teachers, especially the late Donald B. McIntyre, who lit the spark back when plate tectonics was a new idea. —AFG

To Sarah and Autumn, for your support, encouragement, and companionship in the field. —GMS

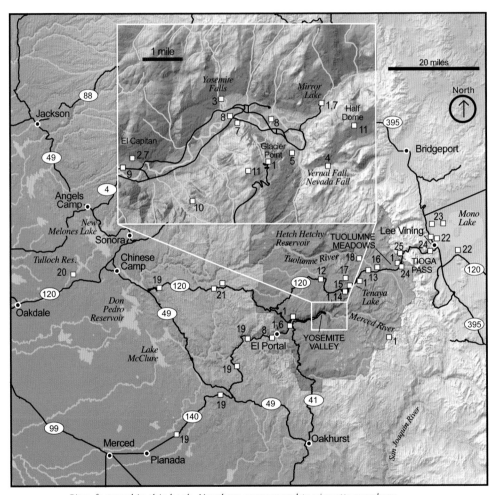

Sites featured in this book. Numbers correspond to vignette numbers.

Contents

Preface

Yosemite National Park is a remarkable place that resonates with people as few other places do. The nearly four million visitors to the park each year testify to this, but the numbers alone don't convey the impact Yosemite has on the human psyche. The first views of Yosemite Valley have brought visitors to tears. People have been known to spontaneously hop on a plane or drive all night for a single day in Yosemite. Whether they are inching up El Capitan or strolling though a grassy meadow, people are drawn to this powerful place. Although some may seek the park's giant sequoias or diverse wildlife, most are captivated by the spectacular scenery, including world-renowned features such as Half Dome, Yosemite Falls, El Capitan, and Vernal Fall. Whether they realize it or not, these people are drawn to Yosemite's geology. Yosemite ranks among the most impressive geological areas on Earth, and geology was the foundation for its creation as a national park in 1890. The famous rocks and landforms of Yosemite have inspired many ideas about how the Earth works, and the park continues to be one of the world's best natural laboratories for geologic research.

Your authors grew up in California and became fascinated early on with its geology. Allen grew up at the foot of the San Gabriel Mountains, began hiking in them while in elementary school, was awakened to the wonders of geology by the 1971 San Fernando earthquake, and spent many long weekends exploring mines in the Mojave Desert. Ironically, he didn't make it to Yosemite until college. Greg grew up in the Sierra Nevada foothills just north of Yosemite and discovered Sierra Nevada geology by exploring its caves and hiking its ridges and canyons. He visited Yosemite often on family camping trips and climbed his first route there as a teenager.

Fittingly, we first met on a geologic field trip to Yosemite in 2002. Greg, then a graduate student, was searching for someone to carpool with and was immediately drawn to Allen, who had wisely rented a convertible; Greg literally jumped into the backseat to claim a spot. This chance encounter led to many years of rewarding discussions and field forays. In the process we discovered, as did geologists Frank Calkins and François Matthes nearly a century before, that combining expertise in petrology (how rocks form) and geomorphology (how landscapes form) can yield interesting insights into the iconic scenery of Yosemite. We hope to share a few of these insights with you in the following pages, but ultimately we'll leave most of the teaching to Yosemite itself. John Muir, one of Yosemite's earliest and most attentive students, perhaps stated it best in 1901 in *Our National Parks*:

> Climb the mountains and get their good tidings. Nature's peace will flow into you as sunshine flows into trees. The winds will blow their own freshness into you, and the storms their energy, while cares will drop off like autumn leaves.

A book of this sort represents a distillation of sources and inspirations too numerous to mention or remember. We would, however, like to acknowledge the help, teachings, and companionship of certain individuals. First and foremost we thank our editor at Mountain Press, James Lainsbury, for his accurate and insightful editing that greatly improved this book. We also thank the late N. King Huber of the U.S. Geological Survey for encouraging us and for providing many stellar examples of effective interpretation of Yosemite's geology. Steve Lipshie, a master guidebook writer and field trip leader, gave the manuscript a thorough review. Eric Knight produced the beautiful landscape illustrations. Scott Bennett drafted most of the maps. Robert Anderson, John Bartley, Steve Bumgardner, Drew Coleman, Brian Collins, Miriam Duhnforth, Scott Hetzler, Eric Knight, Bryan Law, and Cecil Patrick accompanied us on many field excursions and helped tease out many of the stories told here. Former students Walt Gray, Breck Johnson, Ryan Mills, Kent Ratajeski, and Ryan Taylor made their own discoveries in the park and enlightened us along the way.

U.S. Geological Survey scientists Ned Andrews, Paul Bateman, Malcolm Clark, Ron Kistler, Clyde Wahrhaftig, and Gerald Wieczorek, as well as the greater U.S. Geological Survey, have been indispensable in laying out the geology of the Sierra Nevada and making the discoveries that allowed more detailed studies to proceed. Our colleagues in Yosemite National Park, Vickie Mates, Jesse McGahey, Peggy Moore, Jim Roche, Jim Snyder, and Jan Van Wagtendonk, generously supported our efforts in many ways. Academic colleagues Robert Anderson, Tanya Atwater, Bill Bull, Marcus Bursik, Craig Jones, Jessica Lundquist, Steve Martel, Rich Schweickert, Danny Stockli, and John Wakabayashi happily answered questions, posed others, provided data, and generally kept us on track. Fellow climber and geologist Sarah Garlick reviewed the geology of the climbing vignette. The National Science Foundation provided support for several of the studies we have engaged in, a Chapman Family Fellowship gave Allen time off from teaching to work on the book, and the Beck family gave him lodging and meals on many occasions. Greg continues to be inspired by the teachings of his friend Dale Haskamp.

Finally, we express our gratitude to Yosemite for enriching our lives in so many ways.

Introduction

Yosemite National Park and its immediate surroundings display some of the finest geologic features on Earth, and the geology of the park has been the focus of intense study for a century or more. In the simplest terms, Yosemite's geological story consists of the formation of granite and its later sculpting by water and ice, but the grander picture is a tale of ancient supercontinents, the formation and closure of ocean basins, a giant volcanic mountain range similar to today's Andes, San Andreas–type faults, and the repeated appearance and disappearance of massive ice fields and glaciers. We will return to these themes in detail throughout this book but begin here with some important background information.

Yosemite's Geologic Backdrop

Yosemite National Park is an outstanding geological gem in California, a state known for its spectacular and active geology. Although much of California's identity is tied to it being on the West Coast, 1 billion years ago it was just another area in the middle of a continent, perhaps more akin to Kansas than the California we know today. It was near the equator, right in the middle of a supercontinent known as Rodinia—a conglomeration that included all of today's continents. Earth has gone through several cycles during which supercontinents formed and broke up, and Rodinia was the second most recent supercontinent to form.

Changes in the distribution of continents on Earth's surface, and indeed most geologic firepower, such as earthquakes and volcanoes, are a manifestation of plate tectonics. The outer, cooler, more rigid shell of Earth is broken into about a dozen pieces called *plates*. Plates move slowly upon the hotter rocks below them at rates of 1 inch or so per year as even hotter rock wells up from below. This process is like an extremely slow version of the overturn that occurs in a pot of soup placed on a hot stove. Plates consist of the crust, the outermost shell of Earth, and the upper part of the mantle. Earth's basic structure is like that of a soft-boiled egg, with a thin, cool, rigid outer shell (the crust and upper mantle) above the white (the rest of the mantle), which, in turn, sits above the yolk (Earth's core).

The crust is made of rocks that are in large part familiar to most people, such as granite, limestone, sandstone, and gneiss. The mantle, the largest shell of Earth, consists in its shallower parts of rocks rich in olivine, a dense, dark green mineral. Mantle rocks are rarely seen at the surface

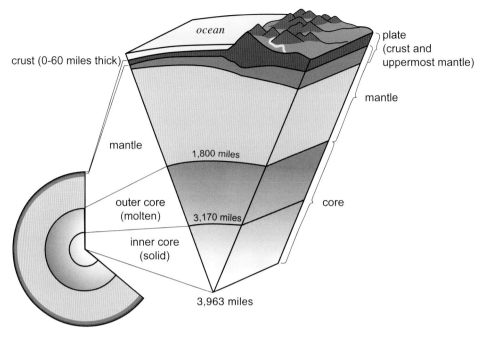

The structure of Earth.

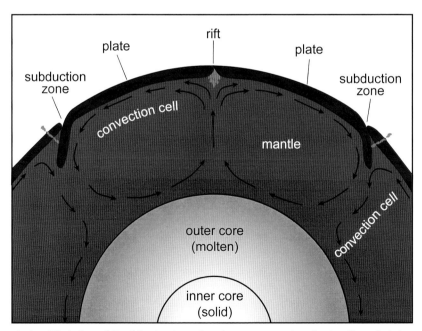

A simplified view of Earth's structure showing our current understanding of plate tectonics. The core is far hotter than the mantle above, so Earth sheds this excess heat by convection in the mantle. These currents move the plates around. Plates are roughly 50 to 100 miles thick and consist of the crust and outermost mantle. Volcanoes are abundant at subduction zones and rifts, which are plate boundaries.

and are prized by geologists because they yield information about Earth's interior. The core, never directly glimpsed, consists of metallic iron and nickel, most of it molten and in motion.

If you were to take a hard-boiled egg, crack the shell into several pieces, and slide those pieces around, you'd quickly discover that there are three basic ways in which plates interact: they can move away from each other, move toward each other, or slide alongside one another. Where plates spread away from one another, volcanoes and ocean basins develop; this type of tectonic margin is known as a *rift*. Where plates slide past one another, great faults occur, such as the San Andreas Fault. Earth's great volcanic chains, such as the Cascade Range, Andes, and Aleutian Islands, form where plates collide and one plate dives under the other. These margins are known as *subduction zones*. All three of these geologic environments have played important roles in the genesis of the Yosemite landscape.

Today the West Coast of North America exhibits all of these plate boundary types. From northernmost California into southern Canada the margin is a subduction zone, where a small plate (the Juan de Fuca Plate) dives beneath the continent, producing the Cascade Range of volcanoes. (Mount St. Helens, which erupted violently in 1980, is part of the Cascade Range.) From northern California south to the Gulf of Mexico the plate margin is the San Andreas Fault, where the great Pacific Plate slides to the northwest against North America, grinding, shaking, and shuddering in great earthquakes. The Gulf of California is a rift, with Baja California, on the Pacific Plate, spreading away from mainland Mexico and forming a new ocean basin.

Understanding Yosemite's geology requires at least an introduction to the depth of geologic time, a concept that even hardened geologists have trouble coming to grips with. Most of Yosemite's bedrock is granite (vignette 1), and most of it formed in a subduction zone about 100 million years ago. This was an unfathomably long time ago if we consider human timescales—about twenty thousand times longer than recorded human history. However, 100 million years represents only about 2 percent of Earth's lifetime, so Yosemite's granites are relative newcomers to the Earth scene.

A simple walk can illustrate the enormity of geologic time—the 4.6 billion years of Earth's history. If each step represents 10 million years, then it takes 460 steps to get from the beginning, when the solar system and Earth formed, to the present. For most people this will be about 400 yards. But let's back up. From the formation of Earth it takes 80 steps (3.8 billion years ago) to get to the age of the oldest known rocks; 340 (1.2 billion years ago) to the first known jellyfish-like animals; 427 (330 million years ago) to the first reptiles; and 454 (60 million years ago) to the first primates. When you have reached 460 steps, draw or imagine a

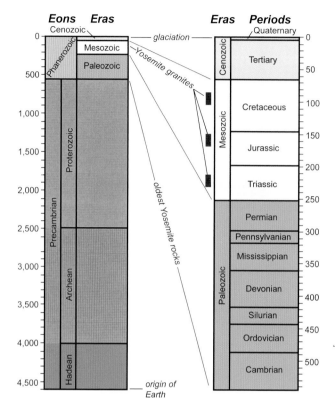

Geologic timescale. Ages are millions of years ago; the right-hand column is an expansion of the last 12 percent of Earth history. Eons, eras, and periods are different subdivisions of geologic time. The red boxes show the major periods of granite formation in Yosemite. Most of Yosemite's granite bedrock formed 105 to 85 million years ago.

pencil line on the ground. The thickness of that pencil line corresponds to the length of recorded human history (about 5,000 years).

Sometime around 800 million years ago Rodinia began rifting apart, breaking into several continents. The pieces that would later become Australia, eastern Antarctica, and southern China rifted away from a large landmass that included most of present-day North America. An equatorial ocean developed between them and steadily widened. After the breakup of Rodinia, California faced this ancient version of the Pacific Ocean known as the Panthalassa Ocean.

The ancestral West Coast of 800 million years ago must have been a dull place. Land life consisted of little more than bacterial films on rocks, and oceans contained only microbes and various jellyfish-like creatures of which little fossil record is left. A vacation to the coast of proto-California would have been boring, and probably smelly as well. Oxygen levels were low, so you would have needed an oxygen mask, and there was little ozone in the upper atmosphere to block ultraviolet rays, so you would have needed to stand in the shade of a tree—except there were no trees.

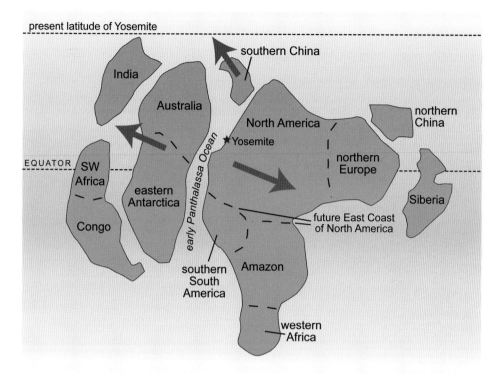

A somewhat speculative reconstruction of the supercontinent Rodinia about 800 million years ago, when it had just begun to rift apart and the Panthalassa Ocean was forming. The arrows show the general movement of some of the fragments. At that time North America sat astride the equator; the future East Coast of North America lay in the middle of the continent, south of the equator, and Yosemite lay to the north. Before rifting, Yosemite lay in the interior of the supercontinent, but it has been at the continental margin, subject to all manner of geologic chaos, ever since. These fragments, and others not shown, dispersed and then gathered again to form another supercontinent, called Pangaea, about 300 million years ago.

The shoreline at that time ran from north-central Utah southwest to the Mojave Desert. It can be traced by following ancient beach sand deposits (now sandstone) that occur in Utah, Arizona, Nevada, and California. The present site of Yosemite National Park was well offshore, where mud was being deposited. This mud became sedimentary rocks that were later metamorphosed. Rocks west and east of the park record this oceanic setting and time period (see vignettes 19 and 25). However, sandy beach deposits are buried in the center of the park at May Lake, where they are far out of place (we'll explore why in vignette 17). The sediments deposited in this post-Rodinia shoreline setting are the oldest known rocks in and around Yosemite.

About 220 million years ago the geologic setting of California changed radically. Subduction replaced rifting off the West Coast as ancestors of the Pacific Plate began descending below the North American Plate. The subduction is recorded by the granites that make up nearly all the bedrock of the park.

Production of magma in subduction zones is a fundamental process that makes Earth different from the other planets in our solar system, and Yosemite has proven to be an excellent place to study it. By chemical and physical processes that are still not well understood (because no one has been deep in a subduction zone), subduction causes rocks deep below the surface of the overriding plate (North America in this case) to melt, forming magma. This liquid rock is less dense than the surrounding mantle and rises buoyantly to either erupt as volcanic rock at Earth's surface or crystallize underground as granite and related rocks. Magma may accumulate underground in magma chambers and then bleed out slowly or come out all at once in colossal eruptions, such as the Krakatoa eruption of 1883 (a large eruption, but far smaller than some that occurred in prehistoric time). A body of unerupted magma that cools and solidifies underground is called a *pluton*, and a large accumulation of plutons is called a *batholith*. Yosemite's bedrock consists of dozens of plutons that are part of the Sierra Nevada Batholith.

Subduction and granite production proceeded in fits and starts from around 220 million years ago until about 85 million years ago, but the majority of the granites in Yosemite National Park and along the spine of the Sierra Nevada are between 105 and 85 million years old. At that time the West Coast was geologically and geographically similar to the modern Andes of South America, with tall volcanic mountains rising steeply from the ocean and a high plateau east of the volcanic chain. These mountains were lifted to high elevations as the compressive tectonic forces of subduction folded the crust and thrust one crustal plate under another, and by the progressive building up of large volcanoes by eruptions. The shoreline had been reoriented from a northeasterly to a northwesterly trend and lay somewhere in the current Central Valley of California.

Although the ancestral Sierra Nevada of 105 to 85 million years ago was undoubtedly a tall, active volcanic range akin to the Andes, the volcanic rocks that lay above the modern Sierra Nevada's granites have long since eroded away. So how do we know that there were volcanoes? First, the midcontinent of North America contains layer after layer of volcanic ash that is the same age as the granites, and there are few other areas that could have been the source. These ashes were produced by eruptions that were far larger than any that have occurred in recorded human history. Second, the sedimentary fill of the Central Valley contains abundant volcanic debris that was eroded off the ancestral Sierra Nevada.

The production of granite below the ancestral Sierra Nevada ceased about 85 million years ago. Why this happened is a mystery, as reconstructions of plate tectonic history suggest that subduction continued for

A reconstruction of what the ancestral Sierra Nevada may have looked like 105 to 85 million years ago, when Yosemite's granites were forming deep below the surface. A long volcanic mountain chain rose steeply from the ocean shore, which was located in today's Central Valley. The mountains probably reached altitudes of 15,000 feet, with glaciers on the higher peaks even during the abnormally warm climate of that time. Some of the volcanoes collapsed during huge eruptions and formed volcanic depressions known as calderas. —Illustration by Eric Knight

tens of millions of years. One hypothesis is that the angle of the subducted plate flattened out, so it no longer reached the depth needed for melting to occur until it got much farther inland. In any event, there is little geologic record preserved in the Sierra Nevada between about 85 and 15 million years ago. This is the time during which erosion took control.

Once granite formation had ceased, erosion began stripping away the overlying volcanic rocks. This was no small feat, considering that the granitic rocks cooled some 2 to 5 miles below the surface. After several tens of millions of years, however, erosion had stripped away most of the volcanic material, exposing the granitic rocks and then carving deeply into them. The eroded sediment was transported westward by river systems draining the range and deposited in the deep subduction trench offshore, which is now the Central Valley. The sediment layers in the Central Valley, also several miles thick, preserve, upside down, most of this erosional history of the ancestral Sierra Nevada.

Volcanism returned to the Sierra Nevada around 15 million years ago when the region north of Yosemite was blanketed with volcanic eruptions similar to those of the present-day Cascade Range. Tall volcanoes were built on granite bedrock, and long lava flows filled many canyons of

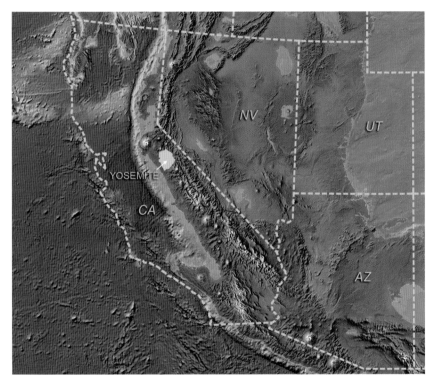

The geographic setting of California around 90 million years ago, at the peak of granite formation. Although the broad outlines are clear, details in this figure are made up, and the figure has not been corrected for the deformation that has occurred since 90 million years ago. The shoreline lay in what is now central California, and the ancestral Sierra Nevada rose abruptly out of the ancestral Pacific Ocean. The deep trench west of the shoreline is now the Central Valley; trenches form in subduction zones, at the boundary where one plate dives under the other. There were probably islands and continental fragments offshore, and a high plateau east of the range that gave way to a vast arm of the ocean that covered the continental interior, including Colorado, much of New Mexico and Utah, and parts of Arizona. Ash from Sierran volcanoes was blown eastward by wind and settled to the bottom of the ocean to form layers that are easily recognizable today in Colorado and other states of the interior continent. —Illustration by Eric Knight, based on a reconstruction by Ron Blakey and Wayne Ranney in *Ancient Landscapes of the Colorado Plateau*

the ancestral Sierra Nevada. These rocks form impressive ramparts along the roads over Sonora, Ebbetts, and Carson passes, including the Stanislaus Table Mountain lava flow (vignette 20), and produce a landscape quite distinct from Yosemite's, with dark, somber, uninspiring summits instead of gleaming granite spires and domes. Volcanism in the past several million years has been focused east of the park, especially around the Mono Basin (see vignettes 22 and 23).

Many clues in the Sierra Nevada landscape suggest that the range experienced a second period of uplift during the past 10 million years. These clues include a distinctly asymmetric shape to the range (a gently dipping west slope and a steeply dipping east slope), active faults along the east slope, tilted volcanic and sedimentary deposits on the west slope (see vignette 20), and narrow bedrock river canyons. Early geologic work suggested that the ancestral, volcanic Sierra Nevada was eroded down to a series of low hills and later uplifted to its present height in the last 10 million years, a rise of many thousands of feet at the crest of the range. It does appear that at least once in the past 10 million years or so the Sierra Nevada Batholith broke along the fault zone bounding the range on the east and tilted westward; however, emerging evidence suggests that prior to this the Sierra was not eroded down to a series of low hills but rather has maintained its stature as a large mountain range since granite production ceased, so the total amount of uplift was probably less than geologists originally thought, perhaps lifting the crest of the range only a few hundred to a few thousand feet. Either way, the uplift caused rivers draining the crest to accelerate their erosion, carving the deep bedrock gorges we see at the bottom of many Sierra river canyons.

A tectonic mechanism for this recent uplift remains elusive and has prompted renewed research into the uplift and erosional history of the Sierra Nevada (your authors are involved in this research), but as often happens, additional research has mostly generated new questions. A clear picture of the late Cenozoic elevation history of the Sierra Nevada has yet to emerge, but it seems that the traditional view of Sierran uplift—substantial erosion followed by substantial uplift—is in need of some revision.

Around 3 million years ago Earth's climate changed significantly, cooling and undergoing large swings in temperature. Why this happened is the subject of much current research, but the record is clear from glacial landforms and from the geochemical signatures that climate change leaves in ocean sediments and ice. Although it is common to hear of "the Ice Age," ice advanced and retreated dozens of times during the last few million years. This pattern is revealed in Yosemite and the rest of the Sierra Nevada and is the subject of several vignettes, especially 9, 13, 14, and 24. To better understand the material in these vignettes, let's learn a bit more about how glaciers work.

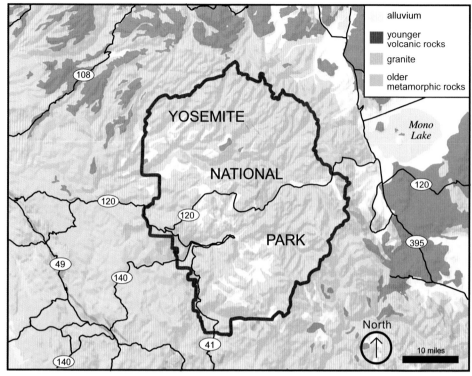

Simplified geologic map of Yosemite National Park and the surrounding region. The park occupies one of the largest unbroken expanses of granite in the Sierra Nevada (Kings Canyon National Park lies in another). The volcanic rocks are significantly younger than the granites and include active volcanoes east of the park in the Mono Basin. Metamorphic rocks include metamorphosed sedimentary and volcanic rocks, many of which are significantly older than the granites. Alluvium includes soil, glacial deposits, and other surface deposits that obscure the bedrock.

How Glaciers Work and How They Shaped Yosemite

When I had scrambled to the top of the moraine, I saw what seemed to be a huge snow-bank, four or five hundred yards in length, by a half a mile in width. Imbedded in its stained and furrowed surface were stones and dirt like that of which the moraine was built. Dirt-stained lines curved across the snow-bank from side to side, and when I observed that these curved lines coincided with the curved moraine, and that the stones and dirt were most abundant near the bottom of the bank, I shouted "A living glacier!"

—John Muir, 1873

With his discovery of a small glacier beneath Merced Peak in October of 1871, John Muir became the first person to verify the presence of glaciers in the Sierra Nevada. At that time, there was disagreement among geologists as to the role that glaciers played in shaping the Yosemite landscape. This discovery proved important in tipping the balance in favor of

Muir and others who argued for a glacial origin of Yosemite Valley and the other iconic landforms of the Sierra Nevada. Indeed, glaciers did play a crucial role in sculpting Yosemite.

In terms of Yosemite viewpoints, none are as famous, or perhaps as spectacular, as Glacier Point. Although a bit crowded in summer (and only accessible by skis in winter), the views across the high country from here are truly breathtaking, and you can peer down a sheer 3,000-foot drop into Yosemite Valley. Visitors to Glacier Point are sometimes confused when they realize they can't see any glaciers. There are, in fact, a few small glaciers clinging to the highest peaks in Yosemite, but they are not visible from here. No one really knows where the name Glacier Point came from, but there is no doubt that this is one of the finest places in the park to view landforms produced by glacial erosion, which we discuss in this section. We apologize in advance for introducing a lot of glacial terminology in this introduction, but these words and phrases, such as *arête*, *tarn*, *hanging valley*, and *cirque*, are both beautiful and useful to those venturing into the mountains.

What Is a Glacier?

What exactly is a glacier? A glacier is a land-based ice mass formed from snow that exists over a period of years and is thick enough to flow downhill under its own weight. Simple, right?

Glaciers are often described as "rivers of ice." This is a useful analogy, as there are some obvious similarities between rivers and glaciers: both are composed of water molecules, begin high in the mountains, and flow downhill in canyons. However, there are also some key differences, and they help illustrate how glaciers work. For starters, rivers usually gain volume as they move downhill, whereas glaciers gain volume as they move down from the mountains until they reach an equilibrium between the accumulation of ice from snow and loss of ice by melting, and beyond that point they shrink. A mountain river typically flows at around 3 feet per second, whereas a glacier may take an entire day or more to move that distance. Rivers are usually not more than a few tens of feet deep, whereas glaciers can be thousands of feet thick. Rivers are not capable of flowing uphill, whereas glaciers are, provided that the surface of the glacier slopes downhill (see vignette 13). Finally, rivers cannot erode below their ultimate base level, which in most cases is sea level. Glaciers, on the other hand, can carve deep canyons well below sea level; these are called *fjords*.

Glaciers are fed by precipitation, as are rivers, but for glaciers this precipitation must be in the form of snow. Much of the Sierra Nevada is blanketed by snow each winter, but almost all of it melts in the summer, feeding rivers. Among the highest peaks, however, winter snow persists through the summer. If snow doesn't melt and accumulates for many years, the feathery snow crystals reconstitute as larger, denser crystals, becoming a type of old, heavy snow known as *firn*. The weight of more snow accumulating on top compacts the firn into solid ice. When ice accumulates to roughly 120 feet thick, it will start to flow under its own weight, becoming a glacier.

The health of a glacier can be described by its mass balance, that is, the balance between input from snowfall and output from melting. The upper end of a glacier is called the *accumulation zone*, for this is where snow accumulates and becomes glacial ice. The lower end is called the *ablation zone*, for this is where glacial ice is lost, or ablated, through melting and sublimation (the conversion of ice and snow directly to water vapor). If a glacier gains a greater mass of ice through snowfall than it loses through melting, the glacier will have a positive mass balance for that year and will advance, or move downslope slightly. If it loses more ice mass through melting than it gains through snowfall, it will have a negative mass balance for that year and will retreat upslope slightly. Thus, the health of a glacier is closely linked to climate (temperature and precipitation), which is why glaciers are hallmarks of climate change, both ancient and modern.

Top: *The view from Glacier Point as it looks today, with Yosemite Falls on the left, Royal Arches and Half Dome in the center, and Vernal and Nevada falls on the right.* Bottom: *The view from Glacier Point as it might have looked during one of the many times glaciers advanced into Yosemite Valley.* —Photo and illustration by Eric Knight, courtesy of the National Park Service

The Lyell Glacier as viewed from the summit of Mt. Maclure. Bare ice on all but the uppermost part of the glacier marks the ablation zone, whereas white snow on the uppermost part marks the accumulation zone. The large crack at the top of the glacier separating the ice from the bedrock is known as a bergschrund, *which forms each year as the glacier moves away from the cirque headwall. At 13,114 feet, Mt. Lyell is the highest peak in Yosemite National Park.*

Glaciers move downslope by two mechanisms: deformation and sliding. Deformation occurs as ice flows downhill due to its weight. To model deformation at home, pour honey on a plate and then tip the plate slightly so the honey flows. The bottom layer of honey sticks to the plate, while the honey above the bottom layer flows over it. In sliding, on the other hand, movement of the glacier over bedrock would be similar to what happens if you first coat the plate with a thin layer of oil; rather than deforming as you tip the plate, the honey slides over the oil.

Scientists have documented these two styles of motion in the field by drilling (usually with steam) a vertical borehole into a glacier and then analyzing changes in the shape and location of the borehole over a period of a year or two. A simple pattern is seen in glaciers around the world. The top of the borehole will have moved farther downslope than the bottom. This downslope bending of the borehole is evidence of ice deformation. As with the honey-on-a-plate example, the bottom of the glacier, which is in contact with bedrock, doesn't move by deformation. The downslope movement by deformation increases upward, with the top of the glacier moving the farthest. One way to visualize this is to set a deck of cards on a table and shear the deck parallel to the table surface with

Your second author feeling the weight of a glacier in a small ice cave at Maclure Glacier's terminus. The grooves in the bottom of the glacier were imprinted on the ice as it slid over the bedrock. Your author is lying on a thin veneer of ice. —Steven Bumgardner photo

your hand. The bottom card should not move, whereas each card above it moves a little farther; the card at the top moves the most because it reflects the cumulative movement of all the cards beneath it. The bottom of the borehole will move too, but not because of deformation. It moves because of the wholesale movement of the glacier due to sliding. Thus, the annual downslope movement of a glacier is a combination of deformation and sliding.

Deformation occurs in any body of ice that is at least 120 feet or so thick. Sliding, however, depends on the presence of liquid water at the bottom of the glacier. Historically, the vast ice sheets in the polar regions have had frozen bases and haven't melted much, and therefore have moved only through deformation. However, recent global warming has caused these ice sheets to begin melting, and thus sliding. Alpine glaciers, such as those that once were numerous in Yosemite, have relatively warm, wet bases (seasonally) and move by both deformation and sliding. Deformation occurs year-round, whereas sliding occurs only when a glacier melts. Since glaciers erode the bedrock beneath them only via the sliding process, glaciers typically work on the bedrock only during the summer.

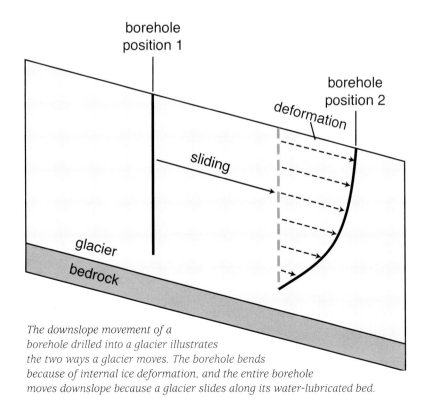

The downslope movement of a borehole drilled into a glacier illustrates the two ways a glacier moves. The borehole bends because of internal ice deformation, and the entire borehole moves downslope because a glacier slides along its water-lubricated bed.

How Glaciers Erode

Glaciers are one of the most effective agents of erosion, although in the past few millennia humans have proven to be more effective. Glaciers typically grind down the bedrock beneath them at rates between 3 and 35 feet per 1,000 years, but erosion rates vary considerably from place to place. The rate depends primarily on the thickness and sliding speed of the glacier, the amount of debris entrained in its base, and the hardness of and joint spacing in (see vignettes 10 and 12) the underlying bedrock.

Glaciers erode the landscape mainly by two mechanisms: abrasion and plucking. Abrasion is analogous to sanding a block of wood with sandpaper. Rocks, sand, and other grit embedded in the base of the ice are the sandpaper that glaciers use to abrade bedrock; ice is far too soft to have any effect by itself. As the embedded rock fragments and the bedrock beneath are ground down, a fine silt called *glacial flour* is produced. When glacial flour is flushed out by meltwater, it often ends up in glacial lakes. It is of such a small particle size that it stays suspended in these lakes for a long time, giving them a distinctive turquoise color that's a telltale sign of glaciers upstream.

The contact between the Maclure Glacier and underlying bedrock. Small rocks embedded in the base of the ice are the tools the glacier uses to abrade and polish the rock over which it slides. The chipped bedrock below the glacier is evidence of plucking. The width of the photo is about 5 feet.

The fine particles produced by abrasion form a layer of grit between the glacier and the bedrock. As the glacier slides, this grit polishes the bedrock. Glacial polish is one of the most obvious marks that glaciers leave on the landscape, and Yosemite is probably the best place in the world to see it. Light reflected from the huge numbers of microscopic scratches on the bedrock surface gives the bedrock its polished appearance. Slightly larger particles embedded in the ice gouge lines into the bedrock that are called *striations*.

Glacial polish and striations are often put forth as evidence of the impressive erosional power of glaciers, but these features really attest to the relative *inability* of glaciers to deeply erode rock. Our studies with Robert Anderson of the University of Colorado indicate that the polish is so impressive in Yosemite, particularly in the area around Tuolumne

Polished and striated bedrock in upper Lyell Canyon. The ice flowed away from the observer. —Josh Helling photo

Meadows and the upper Merced River drainage, because the granite is so hard and free of cracks.

In most glaciated landscapes, plucking (also called *quarrying*) is probably the dominant erosional process, at least when the underlying bedrock is jointed enough to allow it. In order for plucking to occur, there must be water underneath the glacier. The water can originate in two ways: the upper surface of the glacier can melt, with the water traveling down through the glacier in crevasses or moulins (sinkholes in the ice), or the ice can undergo pressure melting when it comes in contact with small bumps on the glacier's bed. Pressure melting is an unusual property of water. For nearly every substance, the temperature at which it melts increases as the pressure increases. Water is different. For ice that is close to its melting point, raising the pressure causes it to melt.

So how does plucking work? Water produced by pressure melting enters the joints in bedrock and refreezes, shattering the rock. The broken fragments are then frozen into the glacier, picked up, and carried away. Although this isn't the only form of plucking, it is probably the most effective because it occurs over wide expanses of the glacier bed and happens continuously as the glacier slides.

In addition to the debris produced by abrasion and plucking, glaciers also transport rockfall debris that falls onto them from canyon walls. The relative proportion of sediment produced by abrasion and plucking versus rockfall in the past is largely unknown, but considering the frequency of rockfalls in Yosemite Valley today, it is reasonable to assume they contributed substantially to the load of sediment Yosemite's glaciers carried.

Glacial Modification of Landscapes

In Yosemite and the rest of the Sierra Nevada, glaciers are not responsible for the gross topography; that work was done earlier by rivers. But glaciers modified these landscapes in dramatic ways, and it is fair to say that without glaciers Yosemite wouldn't contain the famous and iconic landforms that made it a national park. These modifications have created a number of landforms, most of which show that glaciers smooth those landscapes that are completely buried in ice and sharpen those that project above it.

One example of a smoothed landform is a roche moutonnée, which displays the combined effects of abrasion and plucking quite nicely. This landform is a prominent bedrock knob or dome that is gently sloping, smoothed, striated, and polished in places on one flank, and steeply sloping and jagged on the opposing flank. Examples of roches moutonnées in Yosemite include Lembert and Pothole domes, both in the Tuolumne Meadows area. Lembert and Pothole domes differ from the domes on the rim of Yosemite Valley in that they owe their dome shape solely to glacial erosion (see vignettes 11 and 13 for more on the shaping of domes through glaciation and exfoliation).

Landscapes that are not completely buried by ice are sharpened by glaciers. Glaciers on either side of a ridge pluck away rock from its sides until it becomes a narrow knife-edged ridge called an *arête*. Glaciers near the tops of peaks erode cirques, wide, flat-floored amphitheaters with cliff-like headwalls. Glaciers erode cirques headward by plucking rocks from the cirque's headwall and carrying away this rock and rockfall debris. If a peak is surrounded by cirque-forming glaciers, it will become a steep, sharp-edged peak known as a *glacial horn*. The most famous glacial horn is the Matterhorn in Switzerland. Although they aren't as spectacular, Yosemite also has glacial horns, the most famous named—not too surprisingly—Matterhorn Peak.

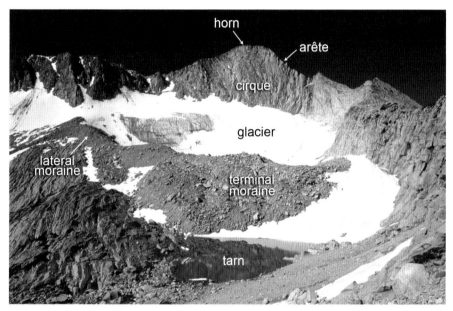

The area around Conness Glacier on the northeast side of Mt. Conness shows numerous glacial features, including a glacier in a cirque, a glacial horn, arêtes, lateral and terminal moraines, and a tarn. Dirty snow marks the paths of rockfalls from the cirque headwall onto the glacier, illustrating one way in which glaciers gather the material that they deposit in moraines.

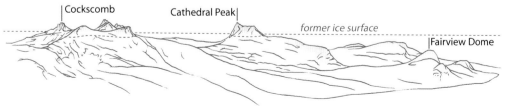

A great example of the contrasting landforms that glaciers produce can be found in Tuolumne Meadows, where sharp peaks such as the Cockscomb and Cathedral Peak, which projected above the tops of the glaciers, tower over smoothed domes such as Fairview Dome, which were overridden by ice.
—Illustration by Eric Knight

One of the most obvious modifications a glacier can make is converting a V-shaped valley carved by a river to a U-shaped valley. Rivers typically carve V-shaped valleys because their erosional power is focused on a relatively narrow channel at the bottom of the valley. In contrast, the erosional power of glaciers extends far up the valley walls. With time, glaciers widen V-shaped valleys, producing U-shaped valleys that have steep walls and relatively flat floors. Yosemite Valley is often said to be a classic example of a U-shaped valley, but that isn't really true. It certainly does have steep walls, but its very flat floor is composed of sediment instead of bedrock; thus, the shape of Yosemite Valley, although influenced by glacial forces, also is a result of the sediment that partially fills it. Lyell Canyon, near Tuolumne Meadows, is a better example of a classic U-shaped valley with a bedrock floor. In many Sierra Nevada valleys and canyons, the farthest down-valley reach of glaciers can be easily discerned by the change from U-shaped to V-shaped cross sections.

Glaciers also produce hanging valleys, which in turn have created the waterfalls for which Yosemite Valley is famous (see vignette 3). Hanging valleys are tributary valleys with floors distinctly higher than those of the main valleys they join. They form because the glaciers that carved out the main valleys, having been much thicker, were able to erode their valleys much deeper than the smaller glaciers in the tributary valleys could. When the glaciers retreated, the tributary valleys were left hanging high up on the sides of the main valleys, and their streams pitch into the main valleys as waterfalls. Bridalveil Fall and Ribbon Fall in Yosemite Valley are classic examples of hanging valley waterfalls.

Another distinctive glacial landform is a tarn. Actually, it is more accurate to say that the tarn occupies the glacial landform. A tarn is a lake that fills a bedrock depression scooped out by a glacier. Unlike rivers, glaciers are able to erode deep basins called *overdeepenings*. As the glaciers retreat, these basins fill with water. Tarns are the most common type of alpine lake, and their distribution in Yosemite matches, almost exactly, the extent of the landscape that was covered by glaciers.

In some unusual cases glaciers excavate particularly large overdeepenings. The best example in the Yosemite region is Yosemite Valley, which is filled with sediment, not water. According to seismic imaging of the Valley, and as confirmed to some extent by wells drilled into the sediments filling the overdeepening, the depth to bedrock beneath the floor of Yosemite Valley is up to 2,000 feet! The depth to bedrock is greatest in the eastern part of the Valley near the Ahwahnee Hotel, shallows near the Three Brothers, and deepens again near El Capitan before shallowing downstream of Pohono Bridge. From here, the Merced River quickly drops 2,000 feet to the town of El Portal, on the western edge of the park. While driving through El Portal, consider that the bedrock floor of Yosemite Valley lies at about the same elevation. Another way to think about

Parker Creek canyon on the east side of the Sierra Nevada south of Mt. Dana displays the U-shaped cross section that is characteristic of glacial erosion. Parker Lake, in the foreground, fills a bedrock depression scooped out by the glacier that occupied the canyon and is also partially dammed by a recessional moraine.

Modification of Yosemite Valley by multiple glaciations.

(A) Yosemite Valley circa 5 million years ago, before any substantial glacial erosion. (B) and (C) Different glacial stages between circa 2 million and 15,000 years ago. (D) Yosemite Valley as it might have looked just after retreat of the Tioga-age glacier 15,000 years ago. (E) Yosemite Valley today. —Illustration by Eric Knight (modified from Matthes, 1930), courtesy of the National Park Service

this overdeepening is to imagine that if all the sediment were washed away by a giant hose, the walls of the Valley would be another 2,000 feet taller!

It isn't clear how this giant hole was excavated. Computer modeling suggests that overdeepenings are the result of increased ice thickness, and hence increased sliding rate, immediately below the junctions of glaciers. The convergence of the Merced and Tenaya glaciers below Half Dome probably focused erosion in this area, scooping out this impressive basin. Given the prodigious amount of sediment fill now occupying this hole, it seems likely that it was not the most recent glaciation that scoured it out, but one of the earlier, much larger glaciations; it likely took several glacial cycles to fill the hole with so much sediment.

What Glaciers Leave Behind

All of the rock debris produced by glaciers is transported down-valley to be left behind in a variety of glacial deposits. The most common glacial deposit is till, jumbled rock debris deposited directly from the ice. Till contains rock fragments of all shapes, compositions, and sizes, ranging from giant boulders to glacial flour.

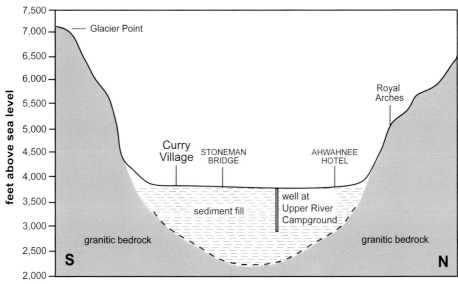

Cross section of eastern Yosemite Valley, where the overdeepening beneath the Valley is deepest. Seismic imaging and wells drilled into Valley sediment suggest that the bedrock is as much as 2,000 feet below the modern Valley floor. Glaciers may have excavated this overdeepening in response to the convergence of the Merced and Tenaya glaciers just up-valley of this location. As the glacier retreated, it deposited till in the overdeepening, and rivers and rockfalls also added sediment and debris. (Modified from Huber, 2007.)

Moraines are ridges composed of till and are also deposited directly from ice without major reworking or transport by running water or wind. The wide range of grain sizes, lack of layering, and mixing of different rock types are diagnostic of moraines and are key pieces of evidence used to distinguish moraines from sediments deposited by a different process, such as stream deposition or rockfall (see vignette 9). In Yosemite, moraines commonly contain a mix of both granitic and metamorphic rock types.

Contrary to popular belief, a glacier does not generally build large moraines by shoving material along in front of it, like a bulldozer. Rather, glaciers pick up rock debris from one place, carry it along, and dump it somewhere else. There is rock debris scattered across the surface of a glacier and embedded in the ice. You might imagine that alpine glaciers look just like icebergs—clear ice with a bluish tint. In fact, alpine glaciers tend to be rather dirty because there is so much rock around to pick up. As previously stated, a glacier can pluck this debris from underlying bedrock, or rockfalls can deposit this debris onto the surface of a glacier.

Glacial till perched on the Diving Board, an overhanging prow of rock on the western shoulder of Half Dome at an elevation of about 7,500 feet. Boulders of Cathedral Peak Granite within the till confirm that it is a glacial deposit, since the rock of the Diving Board is composed of Half Dome Granodiorite. The till's presence on the rim of Yosemite Valley, 3,000 feet above Mirror Lake, indicates that glaciers once filled the eastern side of Yosemite Valley to the rim.

Rocks that fall onto the surface of a glacier can work their way down into the ice, too, as more ice is added to the glacier over time. The debris is eventually dropped when the ice around it melts, forming moraines and other sedimentary deposits.

There are several types of glacial moraines (described in detail in vignettes 9 and 24). Terminal moraines form where rock debris is dumped at the end of a glacier and mark the farthest reach a glacier achieved. Lateral moraines form as rock debris is left along the sides of a glacier. Recessional moraines form as a glacier melts and retreats back up a valley, leaving behind its load of rock debris during occasional pauses in the retreat. The ice does not move backward; rather, the end of the glacier moves up-valley even as the ice flows down-valley because the rate of melting is greater than the rate at which new ice is being added.

Glaciers also leave behind erratics, solitary boulders that a glacier plucks from its bed or that fall onto the glacier, which are then transported downslope by the glacier. As the glacier melts, the boulders are set down onto the bedrock. Erratics from the most recent glaciation are common in Yosemite, especially around Olmsted Point and Tenaya Lake (vignette 14).

History of Glaciers in Yosemite

Many people think the "Ice Age" was a continuous event, and that glaciers were only present once. This is not true. In fact, glaciers advanced and retreated repeatedly across landscapes around the world during Pleistocene time, roughly the past 2 million years. The evidence for repeated glaciations (periods in which the landscape was covered with glacial ice) can be found in Yosemite, but it is incomplete, mainly because each glacial advance wipes out much of the evidence of earlier, smaller glaciations.

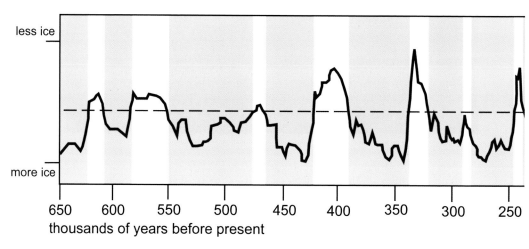

In order to find a complete record of glacial cycles, geologists must turn to deposits found in environments where the record is never reworked. These environments include the ocean floor, lake floors, caves, and glaciers themselves, where various elements archived in sediment and ice provide proxies for past temperatures. Studies of these deposits reveal a rather regular oscillation between warm and cold climates, which caused glaciers to advance and retreat. There were dozens of glacial advances during the last 2 million years, so rather than a singular "Ice Age," the Pleistocene consisted of multiple cycles of glacial and interglacial periods, the glacial portions being "ice ages."

Why Earth more or less regularly cycled through glacial and interglacial climates during Pleistocene time has to do with subtleties of Earth's orbit, which ultimately control the amount of solar radiation Earth (and more specifically, the northern hemisphere, where most of Earth's landmass is) receives from the Sun. The details are beyond the scope of this book, but the crux is that these orbital changes have created a fairly regular pattern of glacial-interglacial cycles, with a major glaciation every 100,000 years or so and minor glaciations about every 20,000 and 41,000 years. When combined, these frequencies produce a sawtooth pattern of warm-cold cycles that closely matches the glacial-interglacial cycles revealed in the deposits mentioned above.

The repeated advance and retreat of glaciers may have been more efficient at eroding the Yosemite landscape than if ice had been present the whole time. As glaciers advanced, they removed the loose and weathered material on the hillslopes. Eventually, they eroded down to more competent rock, where their ability to pluck material away was limited. During interglacial periods, when the glaciers retreated, weathering began again, decomposing Yosemite's granite. This material was then easily carried

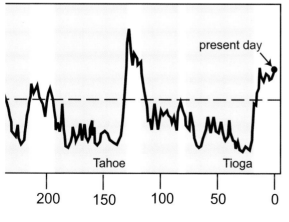

Earth's climate history over the past 650,000 years, derived from an ice core taken from the Antarctic ice sheet. Oxygen-16 and oxygen-18 are different isotopes of oxygen, and their ratio in ice serves as a measure of global temperature and ice volume (see vignette 16). The dashed line separates a generally cold climate (bottom) from a generally warm climate (top). Blue bands represent likely cool and wet glacial periods in the Sierra Nevada, whereas white bands represent warm and dry interglacial periods. The Tioga and Tahoe glaciations are marked.

away during the next glacial advance. The process was not unlike the back-and-forth action of a saw blade on a block of wood, in which chips are first gouged from the wood, then swept away.

Simplified Glacial History of the Sierra Nevada	
(~ means approximately)	
GLACIATION	**TIME FRAME**
Matthes (Little Ice Age)	~AD 1350 to 1850
Tioga	~26,000 to 18,000 years ago
Tahoe	~140,000 to 80,000 years ago
Sherwin	~800,000 years ago
McGee	~1.5 million years ago

The record of early (pre-Tahoe) glaciations in the Sierra Nevada is spotty at best because erosion has removed most of the evidence. There is some evidence suggesting a glacial advance occurred about 1.5 million years ago. It has been dubbed the McGee glaciation because the evidence for it is preserved on McGee Mountain, southeast of the town of Mammoth Lakes. Geologists think the largest glaciation in the Sierra Nevada occurred roughly 800,000 years ago. It is known as the Sherwin glaciation because the best evidence for it is preserved in a roadcut on US 395 near Sherwin Summit (see vignette 23 in *Geology Underfoot in Death Valley and Owens Valley*). In that roadcut the Sherwin Till is found beneath the 760,000-year-old Bishop Tuff, a thick volcanic deposit ejected during a huge eruption of Long Valley Caldera, near the town of Mammoth Lakes. In Yosemite, evidence for the Sherwin glaciation is scarce and appears mainly in the form of isolated erratics and small patches of weathered till. Only in a few locations, such as Turtleback Dome near the western edge of Yosemite Valley, are there verified glacial deposits that are probably of Sherwin age.

The next large glaciation in the Sierra Nevada is called the Tahoe, named for deposits near Lake Tahoe. The timing of the Tahoe glaciation is not entirely clear, but recent dating suggests that it peaked sometime between 140,000 and 80,000 years ago. It was smaller than the Sherwin glaciation before it, but larger than the glaciations that came after. The best place in Yosemite to see Tahoe-age deposits is in Lee Vining Canyon, where huge Tahoe-age lateral moraines can be seen rising above younger Tioga-age lateral moraines (vignette 24).

The most recent large glaciation in the Sierra Nevada occurred between 26,000 and 18,000 years ago during a prolonged cold period. It is known as the Tioga glaciation for glacial deposits near Tioga Pass. The Tioga

This massive boulder on Turtleback Dome near the western end of Yosemite Valley is an erratic that was deposited during an early, Valley-filling glaciation, probably the Sherwin. Both the boulder and the underlying bedrock are El Capitan Granite, but subtle differences distinguish the boulder as an erratic. These include the thin, vertical aplite dike in the boulder (behind and left of the little naturalist), *which doesn't continue into the bedrock underneath, and the greater abundance of enclaves in the bedrock. A weathered cobble of Cathedral Peak Granite in a nearby till deposit confirms that glaciers once reached this spot. The cobble traveled about 16 miles from where the Cathedral Peak Granite is exposed in the upper Merced drainage, and it must have been carried by a glacier to this high spot on the rim of the Valley.*

glaciation is responsible for the majority of glacial features seen in Yosemite today, including moraines, erratics, polish, and striations.

During the height of the Tioga glaciation, a vast ice field formed over and around Tuolumne Meadows. In places it was up to 2,000 feet thick, burying most of the Tuolumne landscape under ice. Glaciers smoothed the landscape under the ice, forming the rounded domes and slabs of Tuolumne Meadows. However, a few peaks and ridges poked up above this sea of ice. Called *nunataks*, these bedrock islands were progressively sharpened as glaciers plucked away at their sides. Classic examples of nunataks in the Tuolumne Meadows area include Cathedral Peak, Unicorn Peak, and Matthes Crest. Larger areas, such as the Dana Plateau and Rancheria Mountain, also became nunataks for short periods as the Tioga glaciers coalesced around them. These nunataks were sanctuaries for a host of plants and animals, ice-free places to which life could retreat as the ice field expanded. A unique type of lupine found on the Dana Plateau probably owes its existence to the fact that the plateau was a nunatak. As the glaciers melted, these species expanded outward from the nunataks to colonize the freshly deglaciated landscape.

It can be difficult to determine the ages of glacial deposits. Until recently, glacial deposits were assigned relative ages based on sharpness of the crests of moraines (moraines become more rounded with time) and the degree of weathering of subsurface boulders. The recent advent of cosmogenic exposure dating (see vignette 9) has allowed geologists to more accurately date moraines, and recent dating in Yosemite tends to confirm what had been the accepted chronology for the glacial-interglacial cycles of the last 2 million years.

During the Tioga glaciation a vast ice field nearly 2,000 feet thick formed in the Tuolumne Meadows region. The landscape below the ice was smoothed by glacial erosion, whereas peaks poking above it were sharpened. Nunataks projecting above the ice field included Cathedral Peak and Matthes Crest. —Illustration by Eric Knight, courtesy of the National Park Service

Modern Glaciers

There are about one hundred small glaciers persisting in the Sierra Nevada today, mostly occupying north- and east-facing cirques at high elevations. Some are in Yosemite, most notably the Lyell and Maclure glaciers. To see either requires a 12-mile hike from Tuolumne Meadows.

The small glaciers in the Sierra Nevada are not the last vestiges of the Tioga glaciation, as is commonly thought. Rather, these glaciers were reborn during the Little Ice Age, a relatively short but cool period that began about AD 1350 and ended about AD 1850; 1850 is also considered to be the approximate date when humans began burning fossil fuels in large quantities. Whether the end of the Little Ice Age was caused by the start of the Industrial Revolution is a subject of much study. The Little Ice Age differs from earlier glaciations in that it is recorded not only in the geologic record, but in the written record as well. In Europe, it affected wine and agricultural production, caused widespread famine, and hastened the outbreak of the bubonic plague. The cold climate doomed the Viking colonies in Greenland, and severe winter weather thwarted Napoleon Bonaparte's invasion of Russia. In Yosemite, the Little Ice Age undoubtedly affected the early inhabitants, but their stories are lost to time. The most obvious relicts of the Little Ice Age in Yosemite are the glaciers clinging to the highest peaks.

The Lyell and Maclure glaciers are the two largest glaciers in Yosemite National Park, and the Lyell Glacier is the second largest in the Sierra Nevada (after the Palisade Glacier). They have retreated rapidly since the peak of the Little Ice Age. Recent work by Hassan Basagic and Andrew Fountain of Portland State University shows that, on average, Sierra Nevada glaciers have lost about 50 percent of their surface area since the end of the Little Ice Age. Some glaciers have transitioned to stagnant ice patches or permanent snowfields (meaning snow that doesn't melt in the summer but isn't thick enough to form ice that flows). Others have been lost entirely. Poignantly, the glacier that John Muir first discovered beneath Merced Peak in October of 1871 is one of those already lost to global warming. If you're interested in seeing a Sierra Nevada glacier firsthand, we recommend that you do it soon.

Glaciers have advanced and retreated across the Yosemite landscape many times in the past, but the recent retreat of the Little Ice Age glaciers is unique in that it is most likely being driven by global warming associated with human activity. Although the long-term climate pattern noted earlier suggests that we should be heading toward another glacial period, the highly elevated levels of greenhouse gases (those that absorb infrared radiation) in our atmosphere may prevent glaciers from returning to Yosemite for a long time.

G. K. Gilbert took the upper photograph of the Lyell Glacier on August 7, 1903 (photo courtesy of the U.S. Geological Survey); Hassan Basagic took the lower photograph on August 14, 2004. The east lobe of the Lyell Glacier (left side) has retreated considerably more than the west lobe (right side), but overall the glacier has lost more than 50 percent of its surface area since about 1880.

RIVERS AND STREAMS IN YOSEMITE

Glaciers played a big role in sculpting Yosemite, but the basic patterns of the landscape had been delineated by rivers and streams much earlier, and these rivers and streams continue to shape the landscape today. If you've been to Yosemite Valley at different times of the year, you know how much the flow of the Merced River varies. In midsummer it is a calm, peaceful river, good for floating and swimming. In fall the flow is but a trickle, and the waterfalls that feed it are mostly dry. But in late winter and spring it roars, fed by snowmelt leaping over gushing waterfalls. This water is important not only for plants, animals, and humans in Yosemite Valley, but for people downstream. The Merced River eventually flows into Lake McClure, an artificial reservoir impounded behind New Exchequer Dam. The reservoir's water is used mostly for irrigation and power generation.

The other large river in Yosemite National Park, the Tuolumne, flows through wilderness to the Hetch Hetchy Reservoir, which has about one-third the capacity of Lake McClure, and on to Don Pedro Reservoir, another artificial reservoir. The water in Hetch Hetchy has a different use: it is

Yosemite Falls at low water in November 2007 (left), *and near peak runoff five months later, in April 2008* (right).

the municipal water supply for the city of San Francisco. Together, these two rivers (including the South Fork Merced River) drain all of Yosemite National Park. A drop of rain falling anywhere in the park will theoretically be directed by topography into one of these two drainage basins, although many drops never actually make it into a river; along the way they fall prey to evaporation, soak into the ground, or are used by plants.

The headwater boundary of the Tuolumne drainage basin is well-defined by the rampart of high peaks (Mt. Gibbs, Mt. Dana, Mt. Conness, Matterhorn Peak, and Tower Peak) that forms the eastern and northern borders of the park, with meltwater from the Lyell and Maclure glaciers forming the source of the Tuolumne River. The high headwaters of the main stem of the Merced River are similarly defined by high peaks along the park's southeastern boundary (Mt. Lyell, Foerster Peak, and Triple Divide Peak) and by Merced Peak and Buena Vista Crest, which separate the main stem from the south fork.

The boundary between the Merced and Tuolumne drainage basins, on the other hand, is curiously subtle. Starting at the park boundary at Mt. Lyell, the divide runs northwestward across Fletcher Peak and Tuolumne Pass through Rafferty and Unicorn peaks, then zigzags west to Echo Peaks and Cathedral Peak. From that point it gets tricky, running near Mt. Hoffmann before turning west across more subtle topography to exit the park near Crane Flat. Along this entire stretch the drainage basin boundary is also the boundary between Mariposa and Tuolumne counties.

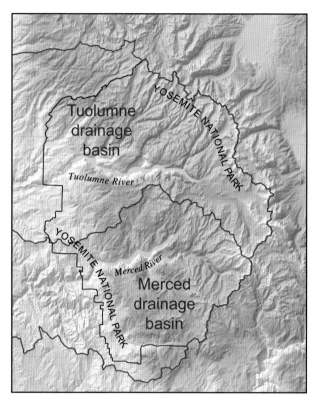

Drainage basins of the Tuolumne and Merced rivers. Both drainage basins are defined by high, sharp peaks on the east, north, and south sides of the park, but the boundary between the two drainage basins is subtle because the Tuolumne ice field ground down the peaks of this divide. The park and drainage basin boundaries coincide along the eastern border of the park.

Why is this divide so convoluted and subtle, whereas the others are high, jagged ridges? Glacial history provides the answer. The northern and eastern ridges were shaped into arêtes by glaciers that flowed down both sides of the ridges, away from one another, whereas the Merced-Tuolumne divide was overrun by several great tongues of ice that spilled over from the vast Tuolumne ice field, which covered Tuolumne Meadows and much of the surrounding region. This one-way flow of ice moved across Tuolumne Meadows and down Tenaya Canyon, smoothing the divide.

When gauging river flow, scientists look particularly at two measures: the average height, called the *stage* of the river (recorded at gaging stations), and discharge. You cannot compare stage from one gaging station to another or from one river to another because the zero point of measurement is arbitrary. The Merced River at Pohono Bridge overtops its banks at a stage of about 10 feet, so flood stage is 10 feet, but at other stations or on other rivers flood stage will be different. There are gaging stations in Yosemite Valley at Happy Isles and at the Pohono Bridge, west of El Capitan. Discharge is usually derived from stage by painstakingly measuring discharge at several times of the year, using flow meters at various depths and positions in the river, and then constructing a graph that relates discharge to stage. Such graphical relations are specific to each gaging station.

Discharge is measured in cubic feet per second (cfs), which is the volume of water passing by a given spot in a second. One cfs is about the flow that a trickling mountain stream produces. That doesn't sound like much, but a cubic foot is about 7.5 gallons, so that's about the amount of water produced by 136 garden hoses going at once, or 648,000 gallons a day. In just one day, a stream flowing at only 1 cfs carries a volume of about 2 acre-feet, enough to cover an acre with 2 feet of water. (An acre-foot is one of those curious English measurements and is equal to about 0.07 bovate-feet or 2.5 oxgang-barleycorns, or, if you prefer, about 5,172 hogsheads. US hogsheads, that is, not UK hogsheads!) And how much is that? A typical North American household uses about 146,000 gallons annually (400 gallons per day), so our trickling stream could supply

Average daily discharge of the Merced River at Pohono Bridge for 2004 and 2006. Several things are clear from this graph:

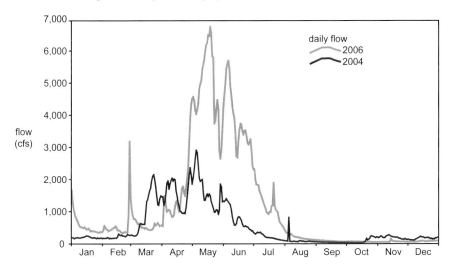

- 2006 was a much wetter year than 2004. Because we're plotting a rate (discharge, or water volume per unit time) against time, the area under the curves estimates the total amount of water that flowed past Pohono Bridge in a given year. These totals were about 35.5 billion cubic feet in 2006 and 15.9 billion cubic feet in 2004. For 2006, that was a total of about 815,000 acre-feet. That's a lot of water, enough for about 1.8 million households.

- Flow is highest in the late spring (usually May); by the first of September the river is very low, awaiting winter storms to freshen up. Late spring (typically May) is usually the best time to view the waterfalls.

- River flow can vary dramatically during short time intervals. For example, average flow on April 24, 2006, was 1,468 cfs, a healthy amount; a week later it was three times higher, at about 4,500 cfs; and by mid-May it was closing in on 7,000 cfs. That high flow would fill a football stadium in a little over an hour.

the water needs of more than 1,600 households. In a year it would fill a medium-sized college football stadium.

The amount of water carried by the Merced River varies dramatically during the year. In late summer and fall the flow drops to around 15 cfs, and in the spring its flow can reach 5,000 cfs or more. During the flood of early 1997 (vignette 8) the river peaked at 26,400 cfs—that's 1,760 times more than the typical low flow in fall.

The streams and rivers of Yosemite contain very chilly water. For example, at the Happy Isles gaging station on the Merced River, the water temperature is close to freezing for much of the winter, climbs to about 50 degrees Fahrenheit by the start of summer, peaks at about 55 to 60 degrees Fahrenheit in midsummer, and then declines back to near-freezing values by the beginning of the next winter. When is the best time to take a dip in the river? Perhaps never, unless you really like being invigorated. However, downstream the water is a bit warmer, especially in summer when flow is lower and the water has more time to be warmed by sun and air as it meanders downstream. Data from 2007 show that the water temperature at Pohono Bridge was about 61 degrees Fahrenheit in June and September and 68 degrees Fahrenheit in July and August, so midsummer is the least bracing time of year to swim in the river.

The huge variations in river discharge, even in nonflood years, mean that stream and river crossings can be dangerous. As of 2005, 154 people had drowned in Yosemite, as documented in *Off the Wall: Death in Yosemite*. By comparison, bears accounted for no deaths during the same period. Bridges and roads must be designed, and in some cases

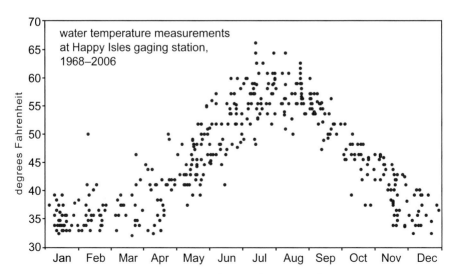

Water temperature measurements taken at the Happy Isles gaging station over a thirty-eight-year period. Hydrologists measure temperature during sporadic field visits. Each dot represents a single temperature measurement.

repeatedly redesigned, to handle these variations. The bridge across Yosemite Creek below Lower Yosemite Fall might seem like overkill in fall, when the creek is but a trickle, but the roaring water of May justifies the sturdiness and height of the bridge. The rivers' erosive power also poses an ongoing threat to park roads; for example, the Merced River downstream of El Capitan had been steadily eating away at sediment supporting California 140 before extensive reconstruction in 2007 and 2008.

Many people feel that the best time to cross a snow-fed stream is in the morning, before the sun has had time to melt snow and send it into creeks and rivers, but the streams themselves feel otherwise. Let's look at some data from the Happy Isles gaging station, this time for hourly discharge on the fifteenth day of each month of 2006:

Variation in discharge at the Happy Isles gaging station on the Merced River plotted by hour of day for the fifteenth day of each month in 2006.

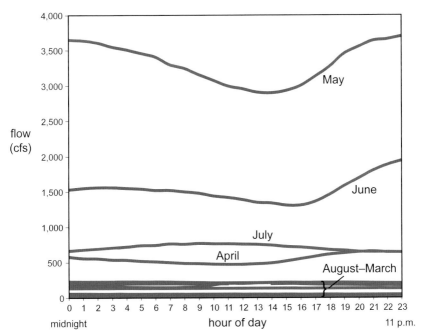

- Flow was much higher on May 15 than in the other months.
- On May 15, the highest flow occurred around midnight (actually 3 a.m. the following morning), and the lowest around 2 p.m. Ditto for April.
- The lowest flow on June 15 occurred around 4 p.m., and the highest was again around midnight.
- Flow peaked during the middle of the day on July 15th.
- Flow was low enough in the other months that it wouldn't matter much when you crossed.

Well, this seems complicated. Let's consider the peak months. Water in the Merced at Happy Isles is the sum of water contributed by all the creeks upstream in the drainage basin. By May the snow line has retreated into higher parts of the park, with little or no local snow to contribute to the runoff. So it makes sense that the peak might come hours later than the warmest part of the day, because the pulse of water generated by snowmelt in the high country takes many hours to make its way downstream to Happy Isles. The water pulse in the Valley is about half a day out of phase with the melt pulse upstream. During wet years, Northside Drive east of El Capitan commonly floods in May during the night, with shallow water lapping across at least one lane, but by midmorning the water has retreated off the road.

So if the best time to cross the Merced River during spring snowmelt is late afternoon, when is the best time to cross little creeks in the high country? Surely that would be the morning, because in the afternoon the snow is melting fastest, and it has only a short distance to go. However, that turns out to be largely false as well. Water melts on the surface of the snow and has to percolate downward through the snow to reach the creeks. This can take a surprisingly long time, up to a day or so. Hydrologists using backcountry sensors to record stream depth can also estimate snow depth because the lag between peak melting in the early afternoon and peak stream flow depends on the distance water has to percolate through the snow. The deeper the snow, the longer the lag.

The debate over which erosive agent—water or ice—sculpts the landscape more effectively over long periods of time rages on. New tools, such as cosmogenic exposure dating, allow better measurement of the rates of these processes, but the answer may never be known.

Plate tectonics, uplift, glaciers, and rivers led to the formation, exposure, and sculpting of the bedrock we see in Yosemite National Park today. A geologic map of the central Sierra Nevada shows that the park occupies one of the largest unbroken expanses of granite in the range. Much of the park's boundary was drawn where granite gives way to other rocks, because it is the gleaming, barren nature of glaciated granite that gives the park its unforgettable scenery. To the west, south, and east, the landscape transitions to somber metamorphic rocks (see vignettes 19 and 25); and to the north it is dominated by dark brown layers of volcanic rocks. These other areas have bears and birds and flowers just as Yosemite does, but they lack the scenic grandeur of the park.

Geologic Study of Yosemite

The spectacular geology of Yosemite National Park has been a focal point for research since the 1860s, when the California Geological Survey was formed. Under the direction of William Brewer, after whom a fine peak in Kings Canyon National Park is named, the survey visited Yosemite in June and July of 1863, making geological, topographical, and botanical observations. Brewer's journal, *Up and Down California*, is an enjoyable and often humorous account of these expeditions and a reminder that it is possible to journey in the mountains without freeze-dried food, high-tech clothing, down sleeping bags, and other modern amenities.

Among the many important contributions the Brewer party made were topographic maps and elevation observations. The latter were made using pairs of mercury barometers, fragile and toxic devices a yard long that were difficult to keep unbroken. One was left at a base camp of known elevation and the other carried to the tops of peaks; the difference in barometric pressure provided a way to measure the difference in elevation. In 1864, Clarence King and James Gardner used barometers and other surveying equipment to make the first accurate shaded relief map of the Valley.

John Muir visited Yosemite in 1868 and took a job tending sheep there in 1869. Among his many brilliant insights, Muir recognized that glaciers had profoundly affected the Yosemite landscape, a concept that had been applied to much of Europe by Louis Agassiz several decades earlier. In this Muir ran afoul of Josiah Whitney, head of the California Geological Survey at the time, who speculated that Yosemite Valley had been produced by a catastrophic earthquake or series of earthquakes that caused the valley floor to drop. Muir's hypothesis won out. Muir was not an infallible observer of nature, however; he proposed that all of California had been covered by glaciers at one time, a hypothesis that is clearly wrong.

Muir's other great contribution to Yosemite was convincing the federal government to set it aside as a national park. Abraham Lincoln had ceded Yosemite Valley and the Mariposa Grove of sequoias to the state of California, but Muir argued for more. His passionate recommendations were followed to some degree by Congress, which set aside a large chunk of land as a national park in 1890 but left Yosemite Valley under state control. To further the conservation effort, Muir and a small group of like-minded people founded the Sierra Club in May of 1892 in the office of San Francisco attorney Warren Olney. Olney and Muir, then fast friends, later split over the issue of damming Hetch Hetchy Valley, a century-old dispute that presaged many environmental battles of today.

Muir lobbied heavily in favor of making Yosemite National Park whole by transferring the Valley and Mariposa Grove to federal control. Well, perhaps *lobbied* is the wrong word; he "campsited" Teddy Roosevelt, taking the famously adventurous president on a backcountry camping trip in 1903 and convincing him around the campfire that consolidating lands into a national park was the best way to preserve the Yosemite landscape. The efforts of Muir and the Sierra Club were rewarded in 1905 when Congress agreed.

Clarence King, one of the young geologists in Brewer's 1863 survey party, went on to become the first director of the U.S. Geological Survey (USGS). This was the beginning of a long and fruitful relationship between the USGS and Yosemite National Park. One of the first projects commissioned by the USGS was a study by Israel Russell, published in 1889, of Mono Basin and the adjacent High Sierra, another name for the higher elevations of the Sierra Nevada. Russell's work presented the first geological descriptions of this now-famous area and included a detailed map of the Lyell Glacier, from which we know the glacier has shrunk in size dramatically in the last century. The much larger Pleistocene version of Mono Lake (vignette 22) was named Lake Russell in Russell's honor.

The debate between Muir and Whitney over the origins of Yosemite Valley lingered unresolved into the twentieth century, so in 1913 the USGS assigned François Matthes to further investigate the issue. Matthes was a topographer by training and therefore had a keen sense of landscape. His detailed studies of the Yosemite landscape, published in 1930, affirmed that glaciers had profoundly influenced the shape of Yosemite Valley but also highlighted the importance of preglacial erosion by rivers. Some of the ideas put forth by Matthes, particularly those regarding uplift of the Sierra Nevada, have since been largely discredited (but remember, he conducted his work well before the theory of plate tectonics had been developed). However, many others have stood the test of time. In particular, his maps showing the extent of ancient glaciers in Yosemite are holding up well in light of recent research. Matthes was greatly aided in his efforts by the bedrock mapping of Frank Calkins, yet another USGS

Left to right: *François Matthes, Sidney Paige, and Frank Calkins setting out on a mapping excursion from their camp near present-day Mather, circa 1916.* —Courtesy of the USGS Photo Library

employee who meticulously studied the rocks in and around Yosemite Valley. Calkins's geologic map of Yosemite Valley remains the definitive map of the Valley's rock distribution.

From the 1940s through the 1980s, USGS geologists, generally led by Paul Bateman and colleagues Dallas Peck and Ron Kistler, systematically mapped the geology of Yosemite, and other USGS scientists mapped much of the rest of the Sierra Nevada. Another USGS geologist, N. King Huber, compiled these maps into a park-wide geologic map in 1987. Huber spent decades studying the geology of Yosemite National Park, and his *Geologic Story of Yosemite National Park* remains one of the best books on the subject. Much of the early work in isotopic methods of dating rocks was performed in Yosemite, and this work carries on to the present as these methods continue to be refined. In the 1980s, Clyde Wahrhaftig of the USGS and University of California at Berkeley revisited the glacier mapping of Matthes and produced the first park-wide glacier map. Spurred by rockfalls triggered by the May 1980 Mammoth Lakes earthquakes, Gerald Wieczorek of the USGS began investigating rockfalls in Yosemite (discussed in vignettes 5, 6, and 7), something he continued to do for the next three decades. Rockfall research in Yosemite is ongoing.

A number of scientists from the USGS, National Park Service, and academia have continued building on the work of these earlier geologists. We have included much of this new research in this book. In light of new findings, some of the old ideas about Yosemite's geologic history are being questioned, or even abandoned, and some of the ideas we present here may turn out to be incorrect. Such is the nature of science. Yosemite National Park has long been a place where new ideas about geology are forged, and that will no doubt continue well into the future.

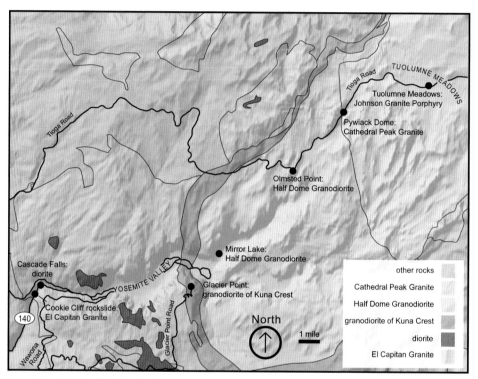

Greatly simplified bedrock geologic map of the Yosemite Valley–Tuolumne Meadows area showing the localities referred to in this vignette; the Tioga Pass locality is not shown. The eastern margin of the main mass of El Capitan Granite is highly complex and includes a unit (Sentinel Granodiorite) not shown on this map. The outcrop of diorite at Cascade Falls is too small to depict.

GETTING THERE

Although rocks are well exposed throughout the park, the best examples of particular types are found in particular places, especially where rockfalls, glaciation, or road construction have cleaned off the surface for you. In this vignette we describe several localities in and around the Valley and along Tioga Road where excellent exposures occur.

1

Bones of the Earth

GRANITE, GRANODIORITE, AND THE BEDROCK OF YOSEMITE

Yosemite's soaring cliffs, deep canyons, glistening domes, and broad uplands are mostly carved out of a rock that can be loosely described as "granite." Although this term is used colloquially for just about any hard rock, especially by the kitchen counter industry, the word *granite* more accurately and specifically refers to a rock composed of visible crystals of quartz, feldspar, and several other minerals. In this vignette we describe what granite and related rocks are and how to recognize them, illustrated by visits to several easy-to-reach spots.

Granite and related rocks form when magma—molten rock—accumulates and crystallizes deep beneath Earth's surface. The molten part of such an accumulation is referred to as a *magma chamber*, and rocks formed from magma are referred to as *igneous rocks*. As magma cools, mineral crystals form and grow, much as ice crystals grow in puddles on cold nights. In fact, ice could be considered a kind of igneous rock because it forms when a liquid cools and crystallizes. The type of igneous rock depends on the type of magma that accumulated and crystallized.

The crystallization of glowing-hot magma is a complicated process. A simple substance such as water crystallizes at one temperature (its freezing point) to a solid (ice) of the same composition, but magmas are chemically complex and crystallize over a wide temperature range, with different types of crystals forming at different temperatures. Granite magma typically begins to crystallize at around 1,700 degrees Fahrenheit (hot enough to glow orange) and finishes crystallization at around 1,200 degrees Fahrenheit. The first crystals to form are those of the minerals hornblende, plagioclase, and biotite, followed at lower temperatures by orthoclase and quartz.

Bodies of granite are called plutons, a word that comes from Pluto, the Roman god of the underworld. Plutons are gods of their own underworld because they are the basic building blocks of Earth's crust, and most other rocks of the crust were ultimately derived from the breakdown of granite. Although plutons are commonly depicted in textbooks as blob-like, many are tabular. Thin sheets of magma injected into cracks are called dikes; in Yosemite, dikes ranging from less than 1 inch to many yards across are common.

The bedrock of Yosemite is almost entirely composed of plutons that crystallized from magma that was injected into the crust between 105 and 85 million years ago. These plutons are complicated because new batches of magma intruded earlier batches throughout this long history. An extensive collection of plutons, such as that found in the Sierra Nevada, is called a *batholith*. Thus, the Sierra Nevada Batholith is composed of plutons that are in turn made up of minerals.

Minerals are naturally occurring substances with a crystalline structure, meaning that their atoms are arranged in defined, symmetrical patterns. Thus, the mineral quartz is composed of silicon and oxygen atoms that are packed in a well-defined symmetrical pattern. In contrast, obsidian, which is chemically similar to quartz, is composed of randomly arranged atoms and is thus not a mineral, but a glass. Color in minerals comes largely from the atoms they contain. Among common minerals, iron generally imparts a strong black or red color.

Five minerals make up most of the granite in the park: quartz, orthoclase, plagioclase, biotite, and hornblende. Although some of these names are formidable, the minerals themselves are easy to recognize and quite friendly once you get to know them, so let's get introduced. A small hand lens of the sort geologists and botanists carry will help you meet them.

First on our list is quartz, a mineral that is familiar to just about everyone. Quartz is silicon dioxide, a union of the elements silicon and oxygen. Quartz is usually clear in isolated crystals, but in rocks it looks light gray

Common Minerals in Yosemite Granites and their Properties

mineral	color	luster	transparency	cleavage	hardness
quartz	clear; may be white or smoky	glassy to slightly greasy	clear	none	7
orthoclase	white, pink	dull or chalky	translucent or opaque	2 at nearly 90°	6
plagioclase	white, gray	dull or chalky	translucent or opaque	2 at nearly 90°	6
biotite	black	shiny	opaque; brown in thin sheets	1; makes thin flakes	2.5
hornblende	black, dark gray, or dark green	glassy to dull	opaque	2; makes diamond-shaped cross sections	5–6
sphene	root beer to honey	shiny	semi-transparent	poor	5–5.5

or slightly brownish gray, with a glassy luster that may appear slightly greasy. It has a hardness of 7 on the Mohs scale, which was created as a standard by which to measure the hardness of different minerals. A knife blade has a hardness of about 5.5, so a knife cannot scratch quartz. Geologists exploit this fact to quickly determine if a white crystal is quartz or calcite (calcium carbonate); calcite has a hardness of 3, so it is easily scratched by a knife. Quartz does not have cleavage, which means that broken surfaces are curved and irregular, like broken glass, rather than flat. The smoky color of some quartz is caused by long-term exposure to radiation from the elements uranium and thorium, which are present in some minerals, such as sphene. Fortunately, these slightly radioactive minerals only occur in trace amounts, so you need not worry about exposure.

Orthoclase and plagioclase are members of the feldspar family of minerals with differing chemical composition. Orthoclase, which contains potassium, is white to pale pink, translucent rather than clear, and has a duller luster than quartz—it sometimes looks chalky rather than shiny. Plagioclase, the calcium-sodium variety of feldspar, is generally white to gray. Both orthoclase and plagioclase have good cleavage, so broken faces are generally flat. These minerals break along two surfaces that are nearly 90 degrees in relation to each other. Typically, fine striations are present on the surface of plagioclase crystals. Geologists rely more on these striations than on color to tell plagioclase and orthoclase apart, because color can tell fibs. A hand lens is quite helpful if you want to see these striations.

Biotite is more chemically complicated than quartz and the feldspars. A black mineral in the mica group, biotite has excellent cleavage that allows it to flake into thin sheets. Biotite crystals are typically hexagonal, bright, and shiny, and can be peeled apart with a fingernail. The dark color is imparted by iron. When biotite crystals weather, they can turn brassy in color; flakes of brassy biotite in streambeds catch the eye because they slightly resemble gold, but gold is much yellower and brighter.

Hornblende is a chemically complex mineral that's rich in calcium, magnesium, iron, aluminum, and silicon. Although hornblende crystals might at first be confused with biotite, they are less shiny and do not form thin flakes. Hornblende crystals typically are black, dark gray, or dark green rectangular shapes on the surface of a rock.

A final mineral that you might see if your eye is sharp is sphene (also known as *titanite*). Sphene has a rich brownish yellow color similar to honey, and its crystals range from microscopic up to 0.25 inch across. Its honey color is unlike any of the other minerals mentioned above. A good place to find sphene is in the Half Dome Granodiorite. Sphene is important because it carries significant quantities of uranium and can thus be used for isotopic dating.

The granitic rocks of Yosemite, and indeed much of the Sierra Nevada, are composed of these minerals in varying proportions. Rocks rich in quartz and the feldspars are light colored and common in the high country of the eastern Sierras; rocks rich in biotite and hornblende are dark colored and more common in the western foothills. Rocks made of different proportions of these minerals are given different names: diorite, granodiorite, and granite, going from darkest to lightest.

The word *granite* comes from an Italian root meaning "grain," as in cereal grain, referring to one of granite's distinguishing characteristics: the mineral crystals are large enough to see, on the order of rice grains or corn kernels. This is the result of slow cooling underground. Had the magma cooled quickly at Earth's surface, most of its crystals would not be large enough to see with the naked eye. Granodiorite and diorite also have large mineral crystals. Technically speaking, most of the rocks in Yosemite are really granodiorite. Granite and granodiorite are distinguished on the basis of the relative proportions of orthoclase and plagioclase. This distinction is not important for our purposes in this book, so we will generally use *granite* to denote the greater granodiorite-granite family of Yosemite's light-colored rocks.

Casual inspection of the walls of Yosemite Valley might lead you to think that all the rocks are medium gray, but this isn't the case. Lichen at these lower elevations obscures the true color of the rocks, as you can tell by looking at the boulders near the valley walls, most of which are covered with a thick coat of gray lichen. The underlying rocks could be any color—lime green, fuchsia, or purple with pink polka dots—and you wouldn't be able to tell without finding a fresh face where the lichen has

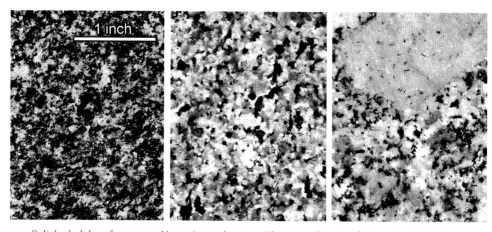

Polished slabs of common Yosemite rock types. The granodiorite of Kuna Crest (left) is a dark rock rich in biotite and hornblende. (Diorite is even darker.) Half Dome Granodiorite (middle) is characterized by black hornblende and biotite, smoky quartz, and white plagioclase and orthoclase. Cathedral Peak Granite (right) is an unmistakable rock characterized by huge crystals of pink orthoclase along with smaller crystals of white plagioclase, black biotite, gray quartz, and pink orthoclase.

been knocked off. At higher elevations, such as around Tuolumne Meadows, lichen is far less abundant and the rocks are much cleaner.

Before we visit Yosemite's rocks, let's clear up a bit of nomenclature. Some rock units have been formally defined and described in the geologic literature, and such units are usually denoted by a place-name followed by a capitalized rock; for example, Half Dome Granodiorite. Others have not been formally described, and such units are given a lowercase rock name in front of a relevant place-name, such as the granodiorite of Kuna Crest. This distinction is irrelevant to most people, but we will strive to keep things in order. And while we're on vocabulary, here's one final point: a group of plutons related in time and space is called an *intrusive suite*, and most of Yosemite National Park is underlain by the Tuolumne Intrusive Suite, which intruded the region between 97 and 85 million years ago.

We begin our tour with a look at the El Capitan Granite, a large unit that makes up much of the western part of Yosemite Valley, as well as most of the cliffs of El Capitan. Boulders of it are beautifully exposed in the Cookie Cliff rockslide, which is discussed in vignette 6.

El Capitan Granite

GETTING THERE
See "Getting There" in vignette 6 for directions.

The beautiful white boulders along the road fell from the cliff to the north in 1982. These boulders by the roadside are El Capitan Granite, a light-colored rock with abundant quartz, plagioclase, orthoclase, and biotite. Both feldspars (orthoclase and plagioclase) are white in the El Capitan Granite, so their relative abundances can only be determined by looking for the striations that characterize plagioclase. Hornblende is scarce in the granite compared to other units in Yosemite, and most of the black minerals are shiny, flaky, hexagonal crystals of biotite.

A more subtle distinguishing characteristic of the El Capitan Granite is the presence of orthoclase crystals that are somewhat larger than the other crystals in the rock—up to 0.5 inch in length. It takes a bit of work to see these larger crystals, and they are sometimes easier to see on a weathered surface than on a fresh one. The sizes of crystals and their arrangement are part of the rock characteristic known as *texture*. Geologists use these characteristics to identify different bodies of rock. It is similar to the way humans have characteristics of a family, such as subtle differences in facial features, stature, hair color, nose shape, and other traits. Although imprecise, this type of rock classification is generally a good guide, and geologists can test the classification of a rock sample in the laboratory by determining its age and chemical composition. Being able to recognize which body of bedrock a boulder came from is useful; for example, it can be used to track the pathways taken by glaciers (see vignettes 9, 14, and 24).

By this time you have undoubtedly noticed all the dark gray blobs of diorite in the much lighter granite. These blobs are technically known by the French word *enclave*, meaning "enclosed." Typically, these pieces of diorite are somewhat flattened, about as big as a fist, and dispersed throughout the granite. Most likely, darker (basaltic) magma was injected into the granite's magma chamber. The two magmas mingled, and the basaltic magma broke into small bits that were dispersed throughout the granite. The enclaves have a high proportion of biotite, which accounts for their dark color.

Diorite enclaves are common in the western part of Yosemite Valley. They were probably derived from the magma that crystallized as the diorite of the Rockslides, a large body of rock that makes up the rubbly slopes west of El Capitan. This large diorite body is mingled with the El Capitan Granite along most of its margins, sometimes in strange and beautiful ways.

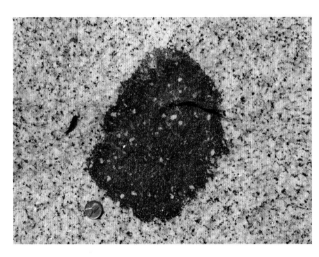

Fresh El Capitan Granite exposed in rubble of the Cookie Cliff rockslide along California 140, with a typical diorite enclave. Note the abundant and obvious white feldspar crystals in the diorite. Penny for scale.

A large boulder of El Capitan Granite in the Cookie Cliff rockslide, with dozens of diorite enclaves.

El Capitan and dark diorite of the Rockslides (rubbly slopes on left) as seen from Tunnel View Overlook along California 140. The diorite has more joints than the El Capitan Granite and thus cannot maintain a steep cliff; note the much more extensive talus beneath the Rockslides.

DIORITE OF THE ROCKSLIDES

GETTING THERE
See the map on page 94.

North of the Cookie Cliff rockslide there is an excellent place to see the diorite and how extensively it is mingled with the El Capitan Granite. To get there from Yosemite Village, follow Northside Drive for 5.7 miles. Stay left (straight) at the junction with Big Oak Flat Road and follow California 140 west (El Portal Road) for 1.8 miles. To get there from El Portal, follow California 140 east (El Portal Road) for 6.4 miles (0.5 mile past the previous locality). Park in the large parking area on the north side of the road at Cascade Creek. Scramble carefully up the boulders along the east (right) side of the creek about 150 to 200 yards to a close view of a large cliff on the left (west) side of the creek. This hike is best done in late summer or fall, when the water is low; do not attempt it during high creek flow or when conditions are icy. The diorite is also exposed extensively along the old Big Oak Flat Road, a decaying dirt road that used to be the main access into the Valley. Look for a pullout on the north side of Northside Drive about 0.25 mile west of El Capitan Meadow. This pullout is where the old Big Oak Flat Road came down to the Valley floor, and you can hike west about 0.5 mile along the old dirt road to sunny exposures of diorite.

Blobs of diorite embedded in El Capitan Granite along the north side of Cascade Creek above California 140. The lowest part of the cliff is mostly granite; above it is a large zone composed mostly of pillow-sized blobs of diorite. This zone and the diorite blobs within it have weathered out in relief from the granite.

Cascade Creek plunges over a series of waterfalls to join the Merced River at this locality, and this hike takes you to the base of one of the cascades. Large blobs of diorite a yard or so across are exposed in the cliff. Notice that on most of the cliff the large, pillowlike diorite enclaves stand out in relief against the granite, whereas closer to the creek they do not and may even form pits. These relative differences in weathering probably result from different microclimates along the side of the creek; areas closer to the creek are wetter owing to splash and spray. Pits and ridges produced by differential weathering provide climbing holds, and many classic climbing routes occur near here.

The El Capitan Granite and the diorite of the Rockslides make up much of the western half of the park and are about 102 million years old. About 95 million years ago magmas of the great Tuolumne Intrusive Suite began to intrude the region east of El Capitan. Among the oldest of these are the granodiorite of Kuna Crest and the granodiorite of Glacier Point. Although these are exposed on opposite sides of the park and have different names, they are related in age and composition; we will refer to them collectively as the granodiorite of Kuna Crest.

The Granodiorite of Kuna Crest

GETTING THERE

The oldest parts of the Tuolumne Intrusive Suite can be examined at Glacier Point or at Tioga Pass. At Glacier Point the granodiorite of Kuna Crest is exposed in outcrops in the parking area and along the walk to the viewpoints. To get there from Yosemite Village, follow Northside Drive for 4.7 miles (from El Portal, follow California 140 east for 9.1 miles), then turn onto Southside Drive and follow it for 1 mile. Turn right onto California 41 south (Wawona Road), drive 9.2 miles, and turn left onto Glacier Point Road. The parking area for the Glacier Point viewpoint is 15 miles down this road. Seeing the equivalent granodiorite at Tioga Pass requires a short but rewarding hike; see "Getting There" for vignette 25. Park on either side of the Tioga Pass Entrance Station and hike about 0.4 mile up and west on the Gaylor Lakes trail to the crest of the ridge. Walk a few hundred yards south (left) to good outcrops of the granodiorite and a great view of the eastern boundary of the park.

The Glacier Point parking area is an especially good place to view the granodiorite because outcrops that were blasted to build the parking area occur at the base of the stairs connecting the lower and upper parking areas. This granodiorite clearly differs from the El Capitan Granite in having more dark minerals (biotite and hornblende), smaller overall crystal size, more uniform crystal size, and a weak alignment of the dark minerals. Big crystals of orthoclase are not present.

Looking east from Glacier Point, the Tuolumne Intrusive Suite makes up the rest of the park, all the way to its eastern boundary. The next unit to the east of the granodiorite here is the massive Half Dome Granodiorite. This unit starts just east of Glacier Point and runs all the way to Tenaya Lake, making up the high ground of Half Dome, Clouds Rest, and several other peaks that are visible from Glacier Point.

Typical rocks of the oldest part of the Tuolumne Intrusive Suite: dark and relatively fine-grained. Penny for scale.

Half Dome Granodiorite

GETTING THERE

The Half Dome Granodiorite is best viewed at Olmsted Point on California 120 (see "Getting There" for vignette 14), and near Mirror Lake in the Valley. At Olmsted Point, park at the viewpoint and walk west to outcrops at the edge of the parking area. Particularly clean, broken boulders of the Half Dome Granodiorite occur just behind these outcrops, downslope from the road.

In the Valley the granodiorite is mostly covered by lichen, but fresh rockfalls near Mirror Lake display it well. To reach one such rockfall (see "Getting There" for vignette 7), follow the paved bike path to where it terminates at the restrooms near Mirror Lake. Continue walking on the wide trail on the northwest side of the sandy area that used to be the lake. After a few hundred yards you will cross a small wooden bridge and then ascend a short series of stone steps. Shortly after the stone steps the trail crosses clean white boulders from several recent rockfalls. The last large boulder on the right side of the trail (walking upstream) is an excellent example of the Half Dome Granodiorite.

The Half Dome Granodiorite is a particularly elegant rock, with large crystals of hornblende, biotite, and both feldspars. Quartz tends to fill in the spaces between the other crystals. A defining characteristic of the Half Dome Granodiorite is the presence of particularly large (over 0.5 inch in length) and well-formed rectangular crystals of hornblende. Biotite forms hexagonal "books" up to 0.25 inch across, and you can easily peel shiny flakes ("pages") of it with a knife blade.

If you want to see what sphene looks like, the Half Dome Granodiorite is the rock to look at. Sphene is easy to find in these rocks, forming honey-colored grains up to the relatively huge size of 0.25 inch across. The crystals are commonly diamond shaped, and the larger ones are truly beautiful, especially when viewed through a hand lens. We have had particular success finding this mineral in the broken rocks at Olmsted Point.

The innermost major unit of the Tuolumne Intrusive Suite is the Cathedral Peak Granite, a spectacular rock with orthoclase crystals that are typically over 1 inch long and sometimes more than 4 inches in length. This rock does not crop out in the Valley, so if you want to see it in place you must go to the Tuolumne Meadows area. However, glacial erratics of the Cathedral Peak Granite do occur here and there in the Valley, most notably along Southside Drive in the Bridalveil Meadow moraine and the El Capitan moraine (vignette 9). These large boulders, more than 10 miles from their homes, are some of the best evidence of the carrying power of glaciers and the paths they followed to lower ground.

Close-up of Half Dome Granodiorite with three crystals of honey-colored sphene. This uranium-bearing mineral is characteristic of the Half Dome Granodiorite. Nickel for scale.

CATHEDRAL PEAK GRANITE

GETTING THERE

Nearly the entire area along Tioga Road between Tenaya Lake and Tioga Pass is underlain by the Cathedral Peak Granite, but roadcuts near Pywiack Dome are the best place to see it. To get there from Big Oak Flat Entrance Station, follow California 120 east (Big Oak Flat Road) for 7.8 miles. Turn left, continuing on California 120 east (Tioga Road) for approximately 32 miles to the beach at the northeast end of Tenaya Lake. From there, drive 1.4 miles to a blasted granite exposure on the right (southeast) side of the road. To get there from Tioga Pass Entrance Station, follow California 120 west (Tioga Road) approximately 13.1 miles and park at similar outcrops on the northwest side of the road.

Orthoclase crystals make up about 25 percent of both the Half Dome Granodiorite and Cathedral Peak Granite, but in the Cathedral Peak Granite most of that 25 percent is concentrated in huge crystals. In the outcrops near Pywiack Dome, blocky orthoclase crystals typically are 3 or 4 inches in length. Such huge crystals are known as *megacrysts*. Many contain ghostly concentric shells composed of minute crystals of biotite, sphene, and plagioclase that apparently were trapped in the orthoclase crystals as they grew. The rest of the rock is composed of plagioclase, quartz, biotite, and smaller crystals of orthoclase.

The most gigantic orthoclase crystals in the Cathedral Peak Granite occur at its contact with the slightly older Half Dome Granodiorite. The outcrops near Pywiack Dome lie in this zone, on the western side of the Cathedral Peak Granite, and the orthoclase crystals in these outcrops are truly gargantuan. Similarly huge crystals occur along the Lyell Fork on the eastern side of the granite body, and searching for that zone makes a fun and easy hike from the Tuolumne Meadows Campground. These crystals are inches longer than the more pedestrian 1- to 2-inch crystals found in the central part of the unit, near the Tuolumne Meadows store.

The Cathedral Peak Granite is unlikely to be mistaken for any of the other rock units in the Yosemite area. However, it could easily be mistaken for the rock that makes up the bulk of Mt. Whitney, 100 miles to the south; the Mono Recesses area, 15 miles southeast of Mammoth Lakes; or the Sonora Pass area, 30 miles to the north. Were these rocks all derived from the same magma body? It's unlikely, given how far apart they are and the long stretches of granite with normal-sized crystals separating them. A more likely conclusion is that Earth has a special recipe for rocks like these that can be reproduced in different kitchens (magma chambers), and when conditions are right, Earth makes them. Geologists

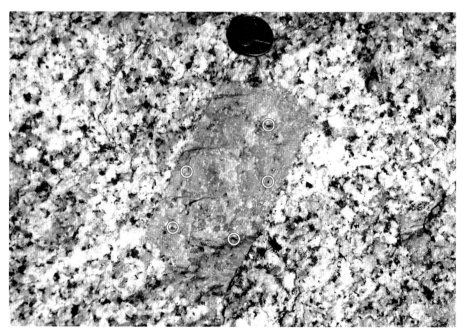

A typical orthoclase crystal of the extra-large variety (a megacryst) exposed in a blasted road-cut in Cathedral Peak Granite along Tioga Road near Pywiack Dome. Circles highlight biotite crystals that are part of a ghostly outline within the crystal. Minerals outside the megacryst include quartz, plagioclase (white), orthoclase (pink), and biotite. Penny for scale.

A glacially polished mosaic of orthoclase megacrysts in Cathedral Peak Granite along the Tuolumne River near Little Devils Postpile (see vignette 18). Knife for scale.

are only just beginning to understand what these conditions are, but very long cooling times (millions of years) and many pulses of thermal energy seem to be required to create such intricate and large crystals.

In many places, especially around Tuolumne Meadows and near Tenaya Lake, the granites are cut by numerous dikes of white, fine-grained rock, ranging from 1 inch to several feet across. Known as *aplite dikes*, they form when a magma body has mostly crystallized and only 10 percent or less of the initial magma is still liquid. This last bit of magma is very fluid and can fill cracks in the rock that has crystallized. Aplite dikes are composed almost entirely of quartz and feldspars. In places, the last bits of liquid crystallized into a rock composed entirely of huge crystals—up to 1 foot or more across. Such dikes are known as *pegmatites*. Aplite dikes and pegmatites are thus granitic rocks distinguished on the basis of their texture.

JOHNSON GRANITE PORPHYRY

One final note: The area around the Tuolumne Meadows Campground is occupied by a rock unit known as the Johnson Granite Porphyry. *Porphyry* is the term for a rock with some larger crystals set in a groundmass of finer crystals, and it is a venerable term in rock nomenclature because

An aplite dike, about 1 foot thick, along the base of Daff Dome, west of Pothole Dome and north of Tioga Road.

The megacryst of orthoclase (center) is part of a rounded reddish blob of Cathedral Peak Granite encased in more typical Johnson Granite Porphyry west of Johnson Peak. One interpretation of this weird rock is that Johnson Granite magma invaded mushy Cathedral Peak magma and entrained blobs of it. Compass for scale.

many ore deposits are related to porphyries. The Johnson Granite Porphyry is indeed a porphyry in many places, but it ranges from aplite to porphyry to pegmatite and other textures that we don't even have names for. In places this unit has a jumbled texture that leads some geologists to suspect that it formed when magma erupted explosively, but no one knows for sure, because volcanic rocks that could be tied to it have not been found; they may have been eroded away.

If you want to see some really weird rocks, hike south from the Tuolumne Meadows Campground on the trail to Elizabeth Lake. Where the trail ends, continue southeast to the large rock slabs between Elizabeth Lake and Johnson Peak, for which this rock unit was named. You will see diverse textural varieties of the Johnson Granite Porphyry, many containing round blobs of what looks like Cathedral Peak Granite surrounding large megacrysts. These are strange rocks indeed.

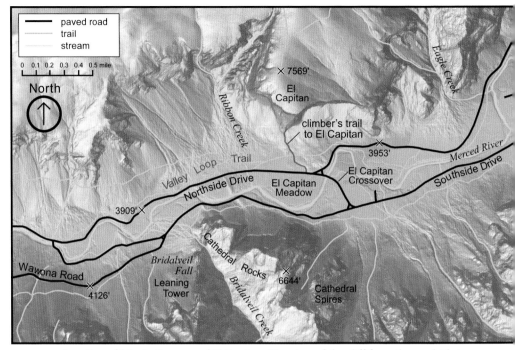

Location of hike to the nose of El Capitan. The view of the cliff from the roadside is outstanding, but be kind to the meadow.

GETTING THERE

We mention several well-known climbing routes and their approximate locations in this vignette, followed by a short hike to the base of El Capitan. To reach El Capitan from Yosemite Village, follow Northside Drive for 2.7 miles to El Capitan Crossover. Drive past this junction and park along the left side of the road adjacent to El Capitan Meadow. To get there from El Portal, follow California 140 east (El Portal Road) for 9.1 miles. Turn right, continuing on Southside Drive for 2.4 miles. Turn left, taking the El Capitan Crossover to Northside Drive and park along the left side of the road adjacent to El Capitan Meadow. Take a minute (or better yet, an hour) to enjoy the view of El Capitan from the meadow, then walk across the road and follow an informal climber's trail toward El Capitan. After a short walk, you will encounter the Valley Loop Trail, which runs along the length of Yosemite Valley. Cross this trail and continue straight on a well-defined climber's trail to the base of El Capitan. As you reach the wall of rock, the trail splits, going both right and left. Take the right fork, which climbs moderately. After about five minutes of walking, you'll come to a flat spot where you can reach out and touch the vast, smooth face of El Capitan.

2

Vertical Exposure

THE GEOLOGY OF YOSEMITE CLIMBING

Early visitors to Yosemite Valley were so impressed by the sheer cliffs that they considered them unclimbable (some modern visitors still do). This sense of impossibility wasn't limited to just tourists; in 1869 Josiah Whitney, then chief of the California Geological Survey, made the following assessment of Half Dome: "[A] perfectly inaccessible [peak] . . . which never has been, and never will be, trodden by human foot." A mere five years later a Scottish sailor named George Anderson succeeding in gaining the summit via the now established route up the northeast side. Rock climbers have now scaled every major rock formation in Yosemite Valley, usually by many different routes. These seemingly unclimbable cliffs are climbable because of the geologic features that adorn them.

Yosemite Valley is considered by many to be the birthplace of modern rock climbing. Just as Mt. Everest inspired mountaineers to achieve new heights, the soaring cliffs of Yosemite Valley inspired rock climbers to push their limits, in terms of what was physically possible and climbing style. Yosemite's cliffs also spurred the development of new equipment that allowed climbers to ascend the cliffs "cleanly," without scarring the rock. Yosemite Valley is where most of the climbing action in the park still takes place, but rock climbing is also popular on the backcountry peaks and along Tioga Road near Tuolumne Meadows. In this vignette we discuss the geologic processes that created the features that allow these amazing walls to be climbed.

Early in the twentieth century, mountaineers developed a classification system to describe the difficulty of climbs, ranging from Class 1 (easy walking) to Class 5 (technical climbing). Class 6 was considered unclimbable without using gear for support and to pull oneself upward. As the rock climbing culture in California flourished in the 1960s, climbers quickly realized that the system was not adequate to describe the new climbs they were establishing. Thus, the Yosemite Decimal System was born, which further subdivides Class 5 climbing. As originally conceived, the system ranged from 5.1 to 5.9, with the next step up from 5.9 being Class 6. As with the original classification system, the Yosemite Decimal System was soon adapted to go beyond 5.9 as climbers inched up ever more challenging walls. At present, the limit is 5.15; this number will probably keep rising, although 5.15 is approaching the limits of human strength and endurance.

Early on, climbers scaled Yosemite Valley's cliffs using pitons (metal spikes) that they pounded into cracks and threaded with slings, which they used to pull themselves upward. This type of climbing, called *aid-climbing*, is still used to ascend many of the big walls in Yosemite Valley. But gradually, as climbers improved their abilities, they developed a new style of climbing that used equipment only to protect against long falls, not to make upward progress. Termed *free-climbing*, this style quickly caught on and is now the most popular kind of climbing in Yosemite. As free-climbing evolved, climbers also noted the damage done to the rock by repeatedly hammering in and removing pitons; what were once thin seams had become wide cracks. New types of equipment, generally called *protection*, were developed that could be placed in existing cracks by the lead climber to protect against a fall and then be easily removed by the following climber, leaving the rock undamaged.

More than most park visitors, rock climbers are intrinsically aware of geology, for their upward progress, and often their very survival, depends on it. Cracks, flakes, dikes, crystals—all of these features allow climbers to excel in the vertical realm. When climbers embark on a big-wall adventure, their lives will be utterly dominated by geology for days on end.

What makes Yosemite so great for rock climbing? Most climbers would agree that it is not a single factor, but rather the remarkable convergence of several factors that makes Yosemite unique. The first has to do with the dominant rock of Yosemite, which of course is granite. Granite is one of the strongest types of rock, due in large part to the slow cooling it underwent and the large interlocking crystals that developed as a result. In addition, most of the granite in Yosemite is exceptionally free of fractures. The sheer vertical face of El Capitan testifies to the great strength and relative lack of fractures in the El Capitan Granite. Climbs in other areas are often on crumbly rock covered with vegetation, but in Yosemite solid, vegetation-free granite is common.

The second geologic factor influencing the quality of climbing in Yosemite is glaciation. Glacial erosion created the vast walls of Yosemite Valley and removed loose, weathered rock, freshening rock surfaces. Glaciers smoothed the landscapes buried under them, sculpting domes such as Pothole Dome in Tuolumne Meadows (vignette 13). For landforms above the ice, the effect was quite the opposite; these landforms were sharpened. For example, ridges that projected above glaciers were steepened as glaciers plucked away blocks from their sides, creating knife-edged ridges called *arêtes*. Arêtes are magnets for rock climbers because of their spectacular scenery and exposed climbing. ("Exposure" here refers to putting oneself in a vastly open vertical space, such as the face of a cliff.) Commonly climbed arêtes in Yosemite include Matthes Crest and the North Ridge of Mt. Conness.

Eichorn Pinnacle, on the shoulder of Cathedral Peak, is an impressive example of a spire that projected above the level of glaciers, sparing its delicate summit from their plucking action.

Glaciers also formed many of Yosemite's spires, such as Eichorn Pinnacle on Cathedral Peak, by carving away the rock below them. Although the earliest of the glaciers that developed in the last 2 million years grew large enough to reach the rim of Yosemite Valley, more recent glaciers, such as those of the Tioga glaciation, did not extend this high; they filled the valley only partway. All of the significant spires in Yosemite Valley sit near the rim because they projected above the level of the more recent glaciers and were spared glacial scouring. This also explains why there are no prominent spires in the canyon of the Tuolumne River: there the recent glaciers did fill the canyon to the rim, removing any spires that may have been present. Examples of spires in Yosemite Valley include the Cathedral Spires east of Cathedral Rocks and the Lost Arrow Spire just east of Upper Yosemite Fall. The dramatic summits of these spires are much sought after by climbers.

Climbing Geologic Features

So, rock type and glaciation together produce the sheer granite walls and spires of Yosemite. However, thanks to numerous smaller-scale geologic features, these seemingly unclimbable walls and spires can in fact be climbed. These features include splitters, flakes, dihedrals, dikes, chickenheads, and megacrysts.

A climber uses the jamming technique to ascend Jamcrack (5.9), a popular splitter just east of Lower Yosemite Fall. The rock on either side of the crack is light colored due to the lack of lichen, which has been worn away by climbers.

Splitters

Yosemite Valley is justifiably famous for the numerous thin cracks splitting its clean granite faces. Called *splitters*, these cracks can be hundreds of feet long and maintain remarkably uniform widths. Many sheer cliffs in Yosemite that appear unclimbable can be climbed with relative ease due to these cracks, which develop in large part due to exfoliation, the process in which shells of rock detach from the main rock mass along exfoliation joints (vignette 12). When it comes to the vertical cliffs of Yosemite, these shells can be hundreds of feet tall and wide. As these vertical slabs of rock are cast off, new rock surfaces are exposed. Once free from the confining pressure of the cast-off slab, the new rock surface expands slightly, producing vertical or curving cracks that split the rock perpendicular to the cliff face. These cracks are usually found splitting pinnacles or rock faces that are bound by other joints that were able to accommodate the expansion. The differences between cracks and joints are subtle; both are kinds of fractures, but joints tend to be more extensive and planar than splitter cracks.

Splitters are climbed by a technique known appropriately as jamming, in which hands and feet are literally jammed into the crack.

Examples of classic splitter climbs in Yosemite include *The Grack* (5.6) on the Glacier Point Apron below Glacier Point, *Reed's Pinnacle Direct* (5.10) above the Big Oak Flat Road, and *Sons of Yesterday* (5.10) near Royal Arches beneath North Dome.

Flakes

Flakes are slabs of rock that are in the process of exfoliating from a cliff. Climbers usually climb flakes by laybacking along their edge: grasping the edge of the flake and placing their feet high on the main rock face, then leaning out from the cliff and climbing hand over hand. Climbing flakes can be a bit unnerving, since they can let go of the cliff face at any moment. Examples of classic flake climbs include *Wheat Thin* (5.10) on Cookie Cliff in the Merced Gorge and *Hermaphrodite Flake* (5.4) on Stately Pleasure Dome above Tenaya Lake. On the famous *Nose Route* of El Capitan, there are two huge and impressive flakes called Boot Flake and Texas Flake. From the valley floor (and even when climbing behind them) it isn't clear what holds these flakes to the cliff face.

A climber ascends the aptly named Boot Flake on the Nose Route *of El Capitan.*
—Tom Evans photo

Experienced big-wall climbers fear the very delicate flakes known as *expanding flakes* because gear inserted behind them can push the flakes away from the wall. This can cause the gear to fall out, or, in the worst case, the flake can peel off the wall. Yosemite climbers have been injured and even killed by expanding flakes.

Dihedrals

Dihedrals are places where two adjacent rock walls come together to form a corner that resembles a partially open book, hence climbers sometimes refer to dihedrals as *open books*. Dihedrals have their origins in rockfalls, a common occurrence along Yosemite's cliffs. When a large slab of rock falls from a cliff face, often detaching along a splitter and an exfoliation joint, the resulting fresh faces form the dihedral. The exfoliation joint that was behind the slab before it fell is almost always exposed at the corner of the dihedral, where it continues into the rock, resulting in a crack that parallels the main cliff face. Climbers ascend dihedrals by bridging between the opposing faces (one foot and one hand on each face), jamming, or laybacking. The dihedrals at the popular climbing area named *Five Open Books*, located west of Lower Yosemite Fall, formed when the huge boulders just west of the Lower Yosemite Fall Trail below the waterfall fell from the cliffs above. Adjacent to Tenaya Lake, and easily visible from the shore of the lake, *Great White Book* (5.6) is another popular dihedral climbing route.

Dikes

Dikes are sheetlike bodies of rock that form when magma is injected into a crack in older rock. Dikes are common in the granitic rocks in Yosemite, where they typically form bands a foot or so thick. For reasons having to do with mineral size and composition, dikes tend to be more resistant to weathering and erosion than the granitic rock they intruded. As a result, dikes often stick out in relief, providing edges and knobs that are perfect for climbing.

Because dikes are not typically associated with cracks into which protection can be placed, many dike routes are notoriously unprotected. Climbers use the term *runout* to describe climbing high above your last piece of protection; for example, if you fall when you are 20 feet above your last piece of protection, you'll drop 40 feet, plus the amount the rope stretches. Thus, runouts are a concern, so climbers add an *R* to a route's rating to indicate moderately runout climbing or an *X* to indicate extremely runout and dangerous climbing. Bolts drilled into the rock often provide the only protection for dike routes, and a total of only two or three bolts per 150-foot pitch is not uncommon. Examples of classic dike climbs in Yosemite include *Snake Dike* (5.7R) on the southwest face of Half Dome (visible with binoculars from Glacier

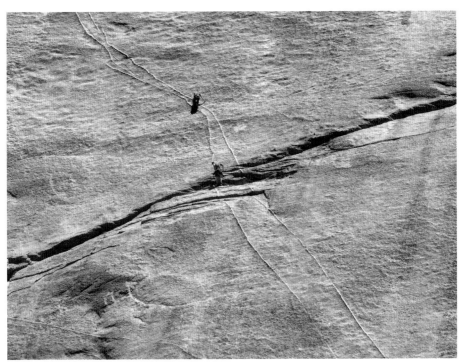

Climbers work their way up a pair of slanting aplite dikes on Dike Route, *Pywiack Dome. The lower climber has an excellent belay stance on a prominent joint. Such comfortable ledges are rare on this runout route.*

Point) and the *Dike Route* (5.9R) on Pywiack Dome near Tenaya Lake (visible from a large gravel turnout on the south side of Tioga Road 0.8 mile east of Tenaya Lake).

Chickenheads

Many granites contain dark blobs of rock that are rich in biotite mica and hornblende. Called *enclaves* (and discussed in vignettes 1 and 6), these blobs are often more resistant to erosion and weathering than the surrounding granite. As a result, they protrude from rock faces. Climbers refer to them as *chickenheads*, perhaps because their shape is reminiscent of a chicken's head (you can decide for yourself). Many are the perfect size for handholds and footholds. Thus, steep and even overhanging walls containing chickenheads can be climbed with "relative" ease. Examples of chickenhead climbs in Yosemite include *Sloth Wall* (5.7), *Boneheads* (5.10), and climbs on the Killer Pillar such as *Fun Terminal* (5.12), all located in the Merced Gorge. This area marks the western edge of the El Capitan Granite, which is characterized by a zone rich in dark enclaves.

A climber makes good use of chickenheads on Killer Pillar at Elephant Rock.

Megacrysts

Megacrysts are exceptionally large crystals. The best examples can be found in the Cathedral Peak Granite, which contains orthoclase crystals that are up to 4 inches or more in length (see vignette 1). Similar to chickenheads, megacrysts tend to be more resistant to weathering and erosion than the surrounding granitic rock, so with time they project out from the rock surface. Where glaciers did not smooth the Cathedral Peak Granite, these megacrysts really project out, sometimes 1 inch or more.

As with chickenheads, megacrysts make excellent handholds and footholds for climbers, albeit relatively small and sometimes breakable ones. Climbers using megacrysts trust their holds to the strength of the crystals, making the runout climbs of Tuolumne Meadows that much more exciting! Many climbing routes in the Cathedral Range south of and around Tuolumne Meadows exploit these megacrysts. In Yosemite, megacryst climbs are limited to exposures of Cathedral Peak Granite. Classic examples include the *Southeast Buttress* (5.6) on Cathedral Peak, which is one of the world's premier alpine climbs, *Golfer's Route* (5.7R) near Tenaya Lake, and the severely runout, exceptionally difficult *Bachar-Yerian Route* (5.11X) near Tuolumne Meadows.

A climber moving up a sea of megacryst knobs on a dome near Tuolumne Meadows.
—Bryan Law photo

Slabs

The large, sloping surface below Glacier Point, dubbed the Glacier Point Apron, is the classic location for slab climbing in Yosemite, although recent rockfalls (see vignette 5) have discouraged climbing there somewhat. Nowhere else in Yosemite Valley is the link between climbing and glaciation so strong. About 20,000 years ago, during the peak of the Tioga glaciation, the top of the glacier in the Valley was likely at about the same elevation as the top of the apron. Below this level the glacier plucked loose rock from the cliffs below Glacier Point. As the glacier receded, it left behind the smooth, low-angle slab that is the Glacier Point Apron. The shallow-angle apron is the lower portion of the U-shape profile characteristic of glacially modified valleys, and there are other aprons like this in Yosemite that present unique climbing challenges. Slab climbing is all about friction: using your body weight to keep your feet stuck to the rock. Modern climbing shoes with sticky rubber soles help a lot. As with dike climbing, areas with slab climbing opportunities are often devoid of cracks or other features in which to place protection, so slab climbers usually rely on bolts for protection and the climbs are often runout. Examples of classic slab climbs on the Glacier Point Apron include *Marginal* (5.9R), *Goodrich Pinnacle* (5.9R), and *The Cow, Left* (5.8R).

Glacier Point. The white arrow marks the approximate height of the Tioga-age glacier. Note how smooth and polished the slabs of the Glacier Point Apron are below the arrow—evidence of glacial scouring.

Boulders and Bouldering

Boulders abound in Yosemite Valley, and the sport of bouldering, which is climbing boulders without ropes or other gear, has become increasingly popular. Perhaps the most impressive boulder in the Valley is the massive Columbia Boulder in the middle of Camp 4. A challenging route on its east face, *Midnight Lightning*, is probably the most famous boulder climb in the world. Tricky boulder climbs are called *boulder problems* because they typically require a climber to figure out an intricate set of moves.

With few exceptions, such as the boulders on the El Capitan and Bridalveil Meadow moraines (discussed in vignette 9), the boulders on the floor of Yosemite Valley are the products of rockfalls. Generally speaking, the larger the boulder, the farther it traveled from the base of the cliff during a rockfall. A large boulder has greater mass than a small one and thus has greater energy for travel, and it's also less likely to get trapped by the ruggedness of the talus slope. Spherical boulders tend to travel farther than those that are angular because they roll more easily. Thus, the largest boulders in Yosemite Valley, such as Columbia Boulder, are typically found near the edge of talus slopes, and even out on the Valley floor, and they tend to be rounded. In the high country of Yosemite, most of the

A climber attempts the difficult boulder problem Midnight Lightning *on the Columbia Boulder in Camp 4. This giant boulder, like most others in Yosemite Valley, owes its existence to a rockfall.* —Bret Meldrum photo

boulders are erratics that were set down by glaciers; great examples can be seen near Olmsted Point and Tenaya Lake (vignette 14). These too are great for bouldering.

GETTING NOSE TO NOSE WITH EL CAPITAN

The geologic factors that make Yosemite unique (strong granite, glaciation, and rockfalls), along with various rock features (splitters, dikes, chickenheads, and so on), have combined to create the exceptional quality of rock climbing in Yosemite, and the spectacular scenery makes the climbing that much better. It's hard to beat climbing a splitter in clean granite on glacially steepened walls, hundreds or even thousands of feet above the Merced River, with the iconic rock formations of Yosemite Valley on the horizon. One climb in particular is a product of all of these geologic factors and features, and it has come to symbolize Yosemite climbing: the *Nose Route* of El Capitan.

Of all of the impressive granite walls in Yosemite Valley, none command the respect and awe of rock climbers as much as El Capitan. Being 3,300 feet high and nearly vertical to overhanging, El Capitan is one of the

El Capitan as seen from the base of Middle Cathedral Rock. The Nose Route *begins at the lowest point of the face and follows the center buttress to the highest point of the face. El Capitan Meadow and Northside Drive are visible in the foreground.*

From the base of El Capitan, one can see the entire 3,000-foot height of the classic Nose Route. *The route (dashed white line) follows the sweep of the left skyline and then takes the far right sunlit cleft to the summit. The two dark spots at the base of the sunlit cleft are roofs that formed when huge blocks of rock fell. The* Nose Route *traverses underneath the roof on the right, called the Great Roof, which overhangs about 15 feet. Climbers on the wall give a sense of the true scale of this gargantuan rock.*

most sought-after big wall climbs in the world. Watching climbers on El Capitan from the meadow below is a favorite pastime of park visitors. If you are visiting the Valley in spring or fall, be sure to see if you can spot climbers on the massive face; they provide an excellent sense of scale. Although it is only a short walk from El Capitan Meadow, few park visitors ever walk to the base of El Capitan (or any other cliff in Yosemite) and put their hands on the rock. That's why we conclude this vignette at the base of El Capitan and the start of the famous *Nose Route*.

Warning! Be advised that anytime you venture near the base of a cliff in Yosemite (or the base of a cliff anywhere, for that matter), you increase your risk of being struck by falling objects. In Yosemite Valley the main hazard is from falling rocks and, in winter, snow and ice, but at the base of El Capitan you are also at risk from gear and sometimes unpleasant items dropped by climbers. Be aware of these hazards, and do not approach the base of El Capitan if you are not willing to accept this risk. Wearing a helmet when at the base is highly recommended.

The first outcrop of rock you encounter on the trail to the base of El Capitan marks the beginning of the *Nose Route*, which is nearly 3,000 feet high. The route was first climbed in 1958. Climbers inched their way up the wall for forty-seven days, hand-drilling holes for bolts that were used as rope anchors. Most climbers now take two to four days to ascend via aid-climbing and some free-climbing, although, incredibly, the *Nose Route* has been climbed completely free, meaning no aid was used for ascent. Even more incredibly, it has been climbed in less than three hours! Depending on the time of year, you may see many teams of climbers working their way up this lofty and famous climb.

Take a minute to put your hands on the rock, lean back, and stare up at the immense wall towering above you—but be careful! More than a few people have toppled over backward trying to take in Yosemite's walls, and your authors have come close. The dominant rock type here is El Capitan Granite, and it is as hard and free of fractures as granites come. The smooth, curving face above you owes most of its shape to the glacial excavation of Yosemite Valley, and if you look carefully you should be able to spot glacial polish on the face above you. Rockfalls from the face have left overhanging edges, called *roofs*. The largest roof, visible from your position, overhangs some 15 feet and is called the Great Roof. On the *Nose Route*, climbers encounter most of the rock features that we described earlier: splitters in the lower section, numerous flakes in the middle section, intrusions and dikes below the Great Roof, and dihedrals above the Great Roof. Without these features, climbing the Nose Route would be much more difficult, and probably a lot less interesting.

There are few granite walls in the world as vast as El Capitan, and probably no others that are so accessible and easily approached. The base of El Capitan is an inspiring place to marvel at the geologic forces that shaped—and continue to shape—this amazing, incredible, and beautiful wall, and also to think about what it takes to climb such a cliff.

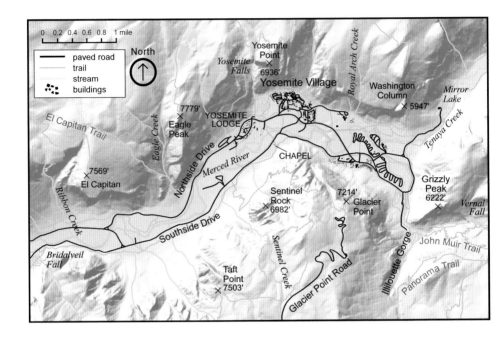

GETTING THERE

Yosemite Falls can be viewed from many spots in the Valley and reached by hikes that range from a short, flat stroll to a strenuous daylong climb. Excellent vantage points are the parking area on the north side of Sentinel Bridge and the open stretch of Southside Drive adjacent to Cook's Meadow between Swinging Bridge and the Chapel. Several easy, short paths from Yosemite Village and Yosemite Lodge lead to Lower Yosemite Fall. The hike to the top of Upper Yosemite Fall begins near Camp 4 and is about 7.2 miles round-trip, with an elevation gain of about 2,700 feet. Plan on a strenuous half day or more. To get to Camp 4, take the shuttle bus or walk from Yosemite Village. If hiking to the top of the falls, consider forging on another mile or so to Yosemite Point, an overlook similar to Glacier Point, high above Yosemite Village with a spectacular view of Lost Arrow Spire only about 100 feet to the southwest.

3
Pushed Off a Cliff

THE ORIGIN OF YOSEMITE FALLS

Next to Half Dome (vignette 11), Yosemite Falls probably is the most iconic image of Yosemite. With a total drop of about 2,425 feet, the three falls make a truly spectacular sight from the Valley floor. It is not the world's tallest, ranking from fifth to merely twentieth in total drop on various lists, but perhaps is the best known, and almost certainly the most photographed, tall waterfall in the world. François Matthes had this to say: "Surpassing all the other falls in height and splendor are the Yosemite Falls . . . They are easily [the Valley's] most spectacular scenic feature. Even more than El Capitan and Half Dome they have given the Yosemite its wide renown."

The flow of water over the falls varies dramatically with the season. In April and May the falls roar with enough water to fill a large football stadium in less than a day, but in late summer and autumn the falls are typically dry, sometimes not even leaving a wet mark at the canyon rim. The volume of water flowing over Yosemite Falls typically peaks at a few thousand cubic feet per second in late spring.

One fascinating aspect of tall waterfalls is the way water dissipates into mist on its way down. At the base of Upper Yosemite Fall, even during high flow, the torrent of water that pours over the lip of the fall spreads out as mist, resembling a veil by the time it reaches the bottom. This is even more clearly displayed at Bridalveil Fall and is the source of that fall's name. Dissipation of the water's force is so thorough that one can stand under the water at times of moderate or low flow without being beaten to a pulp. This is somewhat surprising, given the power of running water (see vignette 8).

A little calculation shows just how much energy is dissipated as the cascades of water turn into mist. If Yosemite Creek is running at 100 cubic feet per second, as it commonly does during the spring, it is pouring nearly 6,300 pounds of water over the falls every second. That's about the weight of two medium-sized cars. Imagine the energy produced by two cars driving over Upper Yosemite Fall every second, falling 1,430 feet, and smashing into the rock. That's a lot of energy, and yet if you stand near the falls you don't feel such a catastrophic energy release. The energy is dissipated as the stream of water is broken into fine droplets

71

Yosemite Falls in full flow, May 2006, viewed from Cook's Meadow. The upper fall drops 1,430 feet, the hidden middle cascade 675 feet, and the lower fall 320 feet. The deep, vegetation-filled ravine to the left of the middle cascade and upper fall once carried the creek's water. Tree-covered benches on either side of the middle cascade formed along massive horizontal joints, and the lower fall drops from the lower bench to the valley floor.

by the drag of the surrounding air and by the wind generated by the falling water. Even though the energy of the falling water is dissipated, we strongly recommend that you use caution when approaching waterfalls in Yosemite. Rocks are also swept over the falls, and their energy is not dissipated!

When the temperature is below freezing, the mist turns to fine ice crystals and piles up at the base of the upper fall in a form given the delightful name *frazil ice*. Ice also commonly coats the rock surrounding the fall at night, and then peels off as the sun hits the rock in the morning. When temperatures remain cold, frazil and fallen ice can accumulate at the base of the upper fall to form an ice cone that typically reaches 100 to 200 feet high, and heights over 300 feet have been recorded.

Yosemite Falls is unique in the park not only for its height, but for its prominence, being in full view from much of the Valley. Most of the park's other waterfalls, such as Nevada, Vernal, Sentinel, and Illilouette, are recessed at least somewhat into side canyons. Upper Yosemite Fall, Sentinel Fall, and Ribbon Fall are unique in leaping off a tall cliff at the very rim of Yosemite Valley, but Sentinel and Ribbon falls carry much less water than Yosemite. Yosemite, Bridalveil, Sentinel, Ribbon, and several other falls are hanging waterfalls, a landform characteristic of formerly glaciated areas. In a typical undisturbed river landscape, tributary streams enter the main river at the level of the river, having worn down their beds to do so. Glaciation dramatically disturbs this nice equilibrium. The larger glaciers in Yosemite Valley cut downward more quickly than the glaciers in tributary valleys, and they also widened Yosemite Valley considerably, leaving the tributary streams hanging high above the main

A remarkably large frazil ice cone at the base of Upper Yosemite Fall dwarfs three climbers, February 1932. —Courtesy of the Yosemite Research Library

Bridalveil Fall viewed from the north rim of the Valley, above the Rockslides near Ribbon Meadow. Note the well-developed valley upstream of the fall. Unlike Yosemite Creek, Bridalveil Creek has been in the same place for a long time—long enough to cut deeply into the bedrock beneath and around it.

Looking north from Sentinel Dome at Yosemite Falls and the gentle upland valley above it. The ravine through which Yosemite Creek used to flow is to the left (west) of the falls.

valley floor once the ice receded. Bridalveil Fall is a particularly clear example of this; from a vantage point across the valley, the broad, steep valley of Bridalveil Creek is clearly truncated. Before glaciation began 2 million years ago, the Valley floor must have been close to the elevation where Bridalveil Creek begins its plunge.

The valley of Yosemite Creek is different. Unlike the deep, broad valley of Bridalveil Creek, Yosemite Creek has carved a shallower notch, about 80 feet deep, in the Valley rim. Why the difference? If you've hiked up the Yosemite Falls Trail, you may have already figured it out. The trail leaves Camp 4 and switchbacks up about 700 feet before heading east and ascending a steep, boulder-strewn ravine via many more switchbacks. This deep ravine seems like a natural place for a creek to flow, and it probably was the former course of Yosemite Creek. But if so, why did the creek decide, at some point, to jump off the cliff instead? A little detective work reveals the answer: the creek did not voluntarily jump off the cliff; it was pushed.

Yosemite Creek's headwaters are on Mt. Hoffmann near May Lake (vignette 17). During the different glacial cycles of the last 2 million years, glaciers formed there, eventually flowing down Yosemite Creek's valley. During the most recent large glaciation, about 20,000 years ago during the Tioga glaciation, the glacier in Yosemite Creek's valley stopped just shy of the lip of the current falls, as shown by many clues, including a conspicuous line of boulders—a small terminal moraine—that the glacier deposited across the creek. Earlier glaciers were almost certainly larger and actually pitched over the rim to join the main glacier in Yosemite Valley. Imagine what a spectacular icefall that must have been!

As the glaciers of the different glacial cycles retreated, they left behind moraines and piles of till in Yosemite Creek's valley. As geologist N. King Huber pointed out, it is likely that these moraines forced Yosemite Creek

This cluster of boulders, located about 0.5 mile upstream of the lip of Upper Yosemite Fall, marks the farthest reach of the Tioga-age glacier in Yosemite Creek.

Left: *Reconstruction of early glaciation in Yosemite Creek's valley and Yosemite Valley, showing what the Yosemite Creek icefall might have looked like from Glacier Point. During the Tioga glaciation, the glacier in Yosemite Creek's valley likely stopped at the rim of Yosemite Valley.* Right: *Yosemite Falls as it looks today from Glacier Point.*
—Illustration by Eric Knight, courtesy of the National Park Service

out of its historic channel, which led to the ravine, and into its current channel to the east, which paralleled the moraines until the creek pitched off the rim of Yosemite Valley. So in essence, Yosemite Creek was pushed out of its bed and off the cliff by glaciers.

So how old is the new dropping-off point of Yosemite Falls? The older moraines upstream of the falls have not yet been dated, so we cannot say for sure. However, the 80-foot depth of the notch at the lip of Upper Yosemite Fall suggests that it has been quite a while. Although the notch is small compared to that of Bridalveil Fall, it is deeper than the notches at the tops of many other waterfalls in Yosemite. For example, the Merced River began cutting the lip of Nevada Fall as the glacier in Yosemite Valley retreated about 15,000 years ago; since that time the bedrock has only been cut a few feet. This suggests that the much deeper notch at the lip of Yosemite Falls must be older, perhaps from the Tahoe glaciation or an even earlier glacial stage. Yosemite Creek was probably diverted from the ravine west of the falls to its present position more than 100,000 years ago. This lucky happenstance of geology produced one of the most spectacular waterfalls in the world.

4

Giant Steps

VERNAL AND NEVADA FALLS

GETTING THERE

See the map on page 70. Vernal Fall is reached by a short but steep hike; Nevada Fall lies beyond Vernal on a more strenuous hike that takes most of a day. The closest trailhead parking area is just east of Curry Village. To get there from El Portal, follow California 140 east (El Portal Road) for 9.1 miles. Turn right, continuing on Southside Drive for 6.1 miles, at which point you will pass the junction with Northside Drive. Continue heading straight, past Curry Village, for another 0.5 mile to the trailhead parking area. From there, you can walk or take the free shuttle bus to Happy Isles. There are many day-use and trailhead parking areas in eastern Yosemite Valley from which you can take the free shuttle bus to Happy Isles to avoid traffic and parking problems.

From Happy Isles, follow the crowd south on the trail, the northern terminus of the John Muir Trail. In about 0.5 mile the trail swings eastward around Grizzly Peak and heads up the gorge of the Merced River. A little less than 1 mile from the start you will cross a bridge to the south side of the river; there is a fine view of Vernal Fall from the bridge. Many people turn around at this point, but the hike beyond is well worth it. Ahead lies the invigorating Mist Trail, with excellent views of Vernal Fall. A few hundred yards past the bridge, the John Muir Trail branches off to the south to climb steeply up to the bench over which Vernal Fall plunges. In early spring this provides a safer way to the top than the Mist Trail, which can be wet and wild (the Mist Trail is closed in winter). The top of Vernal Fall lies about 1.3 miles from the trailhead if you follow the Mist Trail, and a bit farther if you take the John Muir Trail to just above Emerald Pool. The Mist Trail passes the southern side of Emerald Pool for a few tenths of a mile before joining the John Muir Trail again. Eventually, it crosses back to the north side of the river for the climb up to the top of Nevada Fall, which is about 1 mile from the second bridge.

Energetic hikers might want to take a longer, one-way trip to see the falls. You can hike down from Glacier Point along the rim of Illilouette Gorge to Nevada and Vernal falls on the Panorama Trail, and then down to Yosemite Valley, a one-way trip of about 9 miles. An even more strenuous and rewarding loop involves hiking from the valley floor up the Four Mile Trail to Glacier Point, and then down to the falls along the Panorama Trail.

Please note that over twenty people have died when they were swept over these falls. Pay attention and do not enter the water above the falls, even when the water level is low. Several of these deaths occurred when people slipped on rocks while filling bottles, washing their face, or taking photos well above the waterfalls. Also be aware that, because the Mist Trail passes under some very large cliffs, there is a low but ever-present risk of rockfalls. Hike this magnificent trail at your own risk.

Yosemite's waterfalls are perhaps its most notable attractions. As pioneering Yosemite glacial geologist François Matthes said, "Those who view the Yosemite when its falls are dry behold only the empty stage upon which the living waters play their dramatic act." When peak spring runoff approaches, typically in May, visits to the park pick up dramatically, and local news media send reporters to assess the river flows and determine during which weekend the falls will be most splendid.

Waterfalls form when the profile of a streambed or riverbed is interrupted by a discontinuity, which introduces a kink in the otherwise smooth profile. Geologists call these *knickpoints*. Knickpoints can form in many ways. A river may cut through an erosion-resistant layer of rock that overlies a softer layer; the water then erodes the softer rock downstream of the knickpoint, and a waterfall develops where the harder layer hasn't eroded as much. A fault can displace a riverbed, or a mountain range can rise, creating topographical relief. Or a rockfall may block a river's course. Glaciers can also form knickpoints—big ones. In Yosemite, many of the most spectacular waterfalls are hanging waterfalls (see vignette 3) that occur at knickpoints.

Since a knickpoint is a perturbation in the otherwise stable, curved profile that a river seeks to attain, the river system tries to smooth out this perturbation. Erosion is therefore focused at a knickpoint, but smoothing out the perturbation occurs by several different processes and usually takes a lot of time. As a result, a knickpoint will often migrate upstream at rates of inches to tens of feet per thousand years as the river tries to return to a smooth, curved profile. That means waterfalls tend to migrate upstream, too.

Many of the knickpoints of the East Coast formed where resistant metamorphic and igneous rocks give way downstream to softer sedimentary rocks. These sites, sometimes referred to as *fall lines*, were often places where textile mills and other industries were concentrated. Niagara Falls is a knickpoint that formed where the Niagara River cut through a caprock (a relatively durable and impervious layer of rock) of limestone and into underlying softer shale. Many knickpoints in Sierra Nevada river canyons formed along contacts between granitic rocks and harder metamorphic rocks, and where displacement occurred along faults. Others relate to glacial erosion and the lingering effects of uplift.

There are more than forty significant waterfalls in Yosemite Valley, with the exact number depending on how you count them, but many of these dry up once spring runoff has ended. The primary falls are Yosemite, Bridalveil, Vernal, and Nevada. In this vignette we discuss the latter two, which are on the main branch of the Merced River and run with at least a little water all year.

The hike to Vernal Fall begins fairly gently, following the stretch of the Merced River that emerges from a gorge and flows across the floor of Yosemite Valley. Once the trail rounds Grizzly Peak and heads east, though, it gets much steeper, climbing about 260 feet in less than 0.5 mile. The gorge below the trail is choked with huge boulders shed from the cliffs, and the trail crosses several historic rockfall paths. The river drops steeply over this rock debris, forming a series of cascades but no large waterfall. Once past the bridge, you have a distant but still impressive view of Vernal Fall.

Vernal Fall, about 288 feet tall, tumbles over a cliff that strikes northwest at about a 45-degree angle to the westward-flowing river. The trend of this cliff parallels the northeastern cliff that bounds Glacier and Washburn points, as well as many of the joints that define the terrain around

Looking eastward up the Merced River in morning light at Vernal Fall (near) *and Nevada Fall* (far) *during peak runoff in May 2006. The river was flowing at about 3,500 cubic feet per second at Happy Isles, and the mist along the Mist Trail* (center foreground) *was especially thick. The cliffs over which the falls plunge are at right angles to one another.*

Taft Point. The face of Nevada Fall, in contrast, strikes northeast, at a 90-degree angle to the face of Vernal Fall. This trend is parallel to the grand northwest face of Half Dome, the cleft between Liberty Cap and Mt. Broderick just north of the falls, and one of the joint sets around Taft Point. It seems that regional joint patterns (discussed further in vignette 10) control the orientations of the cliffs these falls cascade over.

Now that their orientations are established, we can ask how these waterfalls formed. Are they hanging waterfalls, like Bridalveil? No, they lie along the main branch of the river, not on a tributary to a larger valley. Did they develop where an erosion-resistant layer of rock gave way to softer rock underneath, like at Niagara? No, both lie in homogeneous Half Dome Granodiorite, and there are no obvious differences in rock type that would promote differential weathering.

In fact, geologists have puzzled over these waterfalls for a long time and still have not developed a robust theory for how they formed. In 1930, François Matthes noted that together Vernal and Nevada falls resemble "a giant stairway hewn in granite." He thought their origin was glacial and explained it thusly: "The steps that form Vernal Fall and Nevada Fall owe their prominence to the fact that each is composed of massive granite, which the glacier could only abrade, while the steep front, or 'riser,' is determined by a vertical master joint below which the glacier was able to quarry away successive rock sheets."

Basically, Matthes thought the rock between the two falls and below Vernal Fall was heavily jointed, or weaker, than that which the falls tumble over. The jointing allowed glaciers to erode it more thoroughly, creating the steplike falls. This explanation is consistent with what we can see on the ground, because the cliffs over which the falls pour are indeed massive, relatively unjointed granodiorite, but no one has yet tried to measure the joint spacings to see if the rock between the falls and below Vernal Fall is indeed more jointed. Vernal and Nevada falls both occupy narrow spots in the canyon, whereas the areas upstream and downstream of each are broader. Matthes took this as supporting evidence for his hypothesis, arguing that the glaciers were able to excavate the areas between the resistant steps more easily and broadly, whereas the more resistant rock at the falls resisted sideward erosion.

Field studies of glacial striations indicate that ice can flow across bedrock steps in unexpected ways. Not only did glaciers flow down the steps of the falls, plucking away blocks of rock in the process, ice at the bottom of the glaciers also flowed across and up the faces (risers) of the steps. This type of flow may have enhanced the glacial plucking along the riser, further deepening and steepening the steps.

After crossing the footbridge the trail follows the south side of the Merced River along the cataract below Vernal Fall. There are several excellent vantage points on flat rocks along the trail. At the trail junction a few hundred yards upriver from the bridge, you can follow the Mist Trail along the

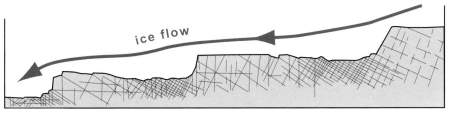

Schematic profile of Matthes's explanation for the formation of a "giant stairway," such as that seen at Vernal and Nevada falls. Steps form where highly jointed granite lies downstream of more massive granite. Glaciers can pluck and carry away blocks from the highly jointed granite, resulting in rapid down-cutting.

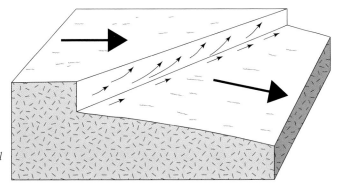

This illustration is modeled after a bedrock step in Glacier National Park, Montana, and it shows how glaciers may have enhanced the bedrock steps at Vernal and Nevada falls. Glacial striations on the bedrock record the direction ice flowed across this step, which is a few feet tall. The overall flow of the glacier was left to right, indicated by the heavy arrows, but striations (light arrows) show that ice also flowed along the base of the riser and up its face. This flow may have enhanced the plucking of bedrock from the riser of the step. (Adapted from R. P. Sharp, Living Ice, *1988.)*

river or the drier John Muir Trail, which switchbacks up the mountain to the south. Both lead to the top of the fall and a spectacular view.

The top of the fall is a wonderful spot and can be quite crowded. Aptly named Emerald Pool lies a short distance above the lip of the fall and is separated from it by a short, rocky cascade. Wading or swimming in the pool is prohibited due to safety concerns. Emerald Pool is fed by a smooth cascade known as Silver Apron. The smooth rock of the apron shows just how unbroken the granite above the fall is, save for a few exfoliation joints.

Above Silver Apron the Mist Trail crosses a sturdy bridge to the north side of the river and traverses a fairly gentle area before ascending a set of switchbacks that climb to the top of Nevada Fall. This climb is below the steep southwestern face of Liberty Cap, which was named for its resemblance to a type of rimless cap that symbolized liberty to ancient Romans and again to revolutionaries during the French and American

Angular boulders near the trail below the steep southwestern face of Liberty Cap. These boulders are the products of rockfalls from Liberty Cap, and this is near the area affected by the 1872 Liberty Cap rockfall (see vignette 7). Nevada Fall is in the background.

View down the face of Nevada Fall, a drop of about 568 feet. Not a view for the acrophobic.

revolutions. The abundance of broken rock here is consistent with Matthes's conjecture that the region between Vernal and Nevada falls is more highly jointed than the rock under the falls, but there are plenty of unbroken slabs, too.

The view from the precipice of Nevada Fall is even more impressive than that from Vernal Fall because the drop is greater, about 568 feet, and because a bridge takes you across the river near the brink. As with Vernal Fall, the rock above the fall is smooth and unbroken except for exfoliation joints. Note how little the Merced River has eroded the lip of the fall during the past 15,000 years or so. (Contrast this with the large notch that Yosemite Creek has created at the lip of Yosemite Falls, discussed in vignette 3.)

After visiting the top of Nevada Fall, you may wish to continue south across the river and loop back along the John Muir Trail to the bridge below Vernal Fall. This path provides striking views of both falls, Liberty Cap, Mt. Broderick, and the surprisingly steep south side of Half Dome.

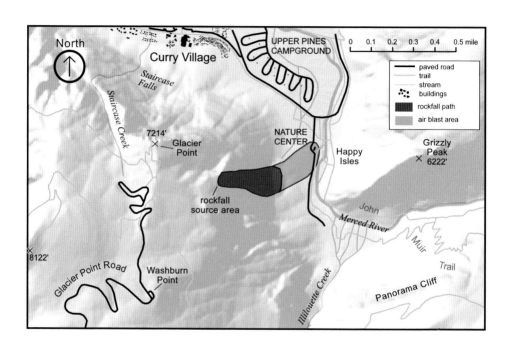

GETTING THERE

The 1996 Happy Isles rockfall is best viewed from the Nature Center at Happy Isles. See "Getting There" in vignette 4 for directions. From the bus stop at Happy Isles, walk south (upriver) for about five minutes along a paved path on the west side of the river to the Nature Center. There are interpretive signs describing the 1996 Happy Isles rockfall and a small viewing area on the south (upriver) side of the Nature Center.

5

Free-Falling Granite

THE 1996 HAPPY ISLES ROCKFALL AND ITS UNUSUAL AIR BLAST

The evening of July 10, 1996, started off like most summer evenings in Yosemite Valley: warm, clear, and calm. Hikers who had reached the summit of Half Dome earlier that day were straggling back to their campsites, and most people were settling down to dinner. Suddenly, at 6:52 p.m., the stillness was shattered by thunderous booms, closely followed by a tremendous rush of air. Dust filled the sky in the Happy Isles area, blotting out the setting sun, and thousands of trees came crashing down. One young man was killed by a falling tree, and many other people were injured, two severely. For minutes those in the area could only see a few feet in front of them because of the choking clouds of dust. A large rockfall had broken loose from Glacier Point.

Rockfalls are common in Yosemite Valley, as they are in most mountainous landscapes, because of the glacially steepened cliffs, fractured rocks, and climatic setting (severe winter storms and freezing temperatures can trigger rockfalls, as discussed below). There have been more than six hundred recorded rockfalls in Yosemite since 1857, the vast majority of which occurred in Yosemite Valley, and there were doubtless many more that were not recorded. In addition, virtually every cliff base in Yosemite Valley is heaped with piles of rockfall debris (talus) that testify to the great number of rockfalls that have occurred since the Tioga-age glacier retreated from the Valley about 15,000 years ago. Yet as common as rockfalls are in Yosemite Valley, the 1996 Happy Isles rockfall was unusual.

The 1996 Happy Isles rockfall actually consisted of two large rockfalls, separated in time by about 14 seconds. The combined volume of these rockfalls was about 40,000 cubic yards, a volume roughly equivalent to a slab of rock about 18 feet thick and the size of a football field. The mass of such a slab would be nearly 90,000 tons. Both rock masses originated from a curving arch near the rim of Yosemite Valley southeast of Glacier Point and directly above Happy Isles. After the rock masses detached, they fell onto an inclined surface and slid for about 800 feet. This inclined surface acted as a launching ramp, pitching the rocks out into a free fall. Still mostly intact, the rock masses then accelerated rapidly for roughly 1,800 feet and hit the talus slope at the base of the cliff at speeds of about

270 miles per hour. The resulting shock wave was recorded by seismometers over 120 miles away and produced local ground shaking similar to that of a magnitude 2.1 earthquake.

The impact area of the rockfall was not particularly large, but because the rocks were in a free fall and hit the ground as intact blocks, their impact generated an air blast that proved to be much more devastating than the falling rocks. Imagine dropping a thick book onto a table covered with loose papers. As the book approaches the table surface, it traps air underneath it and forces the air outward, blowing the papers off the table. Now imagine our football-field-sized slab hitting the ground near you at Happy Isles. Wind speeds near the point of impact exceeded 250 miles per hour, about 100 miles per hour greater than the threshold for a Category 5 hurricane. More than one thousand trees at Happy Isles were either toppled completely or snapped off by the force of the blast. Even trees 300 feet away were toppled, snapped, or stripped of bark. It was these falling trees, rather than rock debris, that caused the fatality and the

The rockfall source area, travel path of rock debris, impact area, and air blast area of the 1996 Happy Isles rockfall as seen from Sierra Point. In this 2006 photo, vegetation is starting to recolonize the impact and air blast areas. The rockfall originated from near the rim of Yosemite Valley, above the approximate upper limit of the Tioga glaciation, where the rock is more weathered. The smoothness of the rock in the bottom two-thirds of the cliff indicates that it was scoured by a glacier.

The dust cloud generated by the 1996 Happy Isles rockfall and air blast. The Nature Center at Happy Isles lies in the center of the expanding dust cloud. The photographer was climbing near Royal Arches at the time. —David F. Walter photo

injuries. The displaced air was thick with dust and small rock fragments, and some rock fragments were embedded in tree trunks over 500 feet away from the point of impact. During the night, smaller sections of rock fell from the same area.

From the viewpoint near the Nature Center it isn't easy to see the talus slope anymore because trees have started to recolonize the area. However, you can still see the trunks of large conifers that were sheared off by the air blast sticking up above the new vegetation. Interestingly, before the 1996 rockfall the park botanist at the time considered Happy Isles to be one of the safer areas in regard to rockfalls because the trees there were much larger than elsewhere in Yosemite Valley. She reasoned that in order for the trees to have grown so large rockfalls must be rare. This is a good example of how thinking on different timescales can alter our perception of geologic hazards. Certainly the air blast at Happy Isles was unusual and particularly devastating, but even a cursory exploration of the talus slope beneath Glacier Point reveals huge amounts of talus, including one large rock avalanche deposit (see vignette 7). Huge boulders tens of feet on a side sitting along the Merced River just upstream of the Nature Center testify to the very large rockfalls that originated from Glacier Point in prehistoric time. The forest at Happy Isles may have had

This huge boulder near the Nature Center is evidence of a prehistoric rockfall at Glacier Point.

some of the largest trees in Yosemite Valley, but even their age was only a blink of an eye in geologic time. When viewed from the geologic perspective, rockfall activity at Glacier Point, and in Yosemite Valley as a whole, has been essentially constant.

With the retreat of the Tioga-age glacier, rockfalls became the primary sculptors of Yosemite Valley. Based on documentation of historic rockfalls, and on calculations of talus volumes, thousands of tons of rock tumble into Yosemite Valley each year on average. Not even the mighty Merced River in flood can compete with that.

The largest rockfall in Yosemite in historic time fell from high on Middle Brother, between Yosemite Falls and El Capitan, on March 10, 1987. After several hours of small rockfalls, a huge portion of the face of Middle Brother let loose and fell some 2,600 feet to the Valley floor. Because the rock mass was not free falling, it broke up substantially on the way down and did not produce a large air blast. It did produce a huge dust cloud, though, and boulders rained down on Northside Drive, some even landing on the other side of the Merced River. The estimated volume of the rockfall was roughly 785,000 cubic yards, or some 1.8 million tons! That's about fifty thousand dump truck loads, and twenty times larger than the 1996 Happy Isles rockfall. Fortunately, park officials had closed Northside Drive shortly after the smaller rockfalls started, so there were no injuries. However, several hundred feet of Northside Drive lay buried under more than 10 feet of rock, and it was two months before the road was reopened. The 1982 Cookie Cliff rockslide (vignette 6) and the 2006 Ferguson rockslide (see vignette 19) provide other examples of the impact geologic events can have on human infrastructure.

Why do rockfalls happen? The simple answer is gravity, but beyond that the explanation is more complex. Geologists begin by differentiating between rockfall causes and triggers. Causes include rock type, degree of weathering, spacing and orientation of fractures, and other factors that affect the long-term stability of rock masses. Triggers are the specific forces or events that initiate the failure. Rock type and degree of weathering both probably played a role in weakening the rocks that ultimately fell in the 1996 Happy Isles rockfall. Although composed of relatively strong granite, the rockfall source area is well above the upper limit reached by

Large talus piles have been accumulating at the base of Yosemite Valley's cliffs since the Tioga-age glacier retreated from the Valley. Middle Brother, shown here, has been one of the most active rockfall areas in Yosemite Valley and was the site of the largest historical rockfall in Yosemite, a 1.8-million-ton failure on March 10, 1987. Though large by Yosemite Valley standards, this rockfall only added slightly to the huge talus pile in the lower left corner of the photo. Light-colored areas and treeless talus slopes mark places where rocks recently detached from the cliffs. The falling rocks removed the stained surface rock and scoured the slopes beneath of trees and other vegetation.

the Tioga-age glacier in Yosemite Valley. The glacier scraped off loose, weathered rock below this level, leaving smooth slabs of unweathered rock popular with climbers (see vignette 2). However, the glacier didn't touch the rocks above this level, so they are more deeply weathered and therefore more prone to failure. This is why the largest rockfalls in Yosemite Valley tend to originate near its rim.

Once a rock is destabilized, it continues to cling to a cliff as long as the driving force of gravity is balanced by the resisting forces of friction and cohesion. A number of forces can upset this balance, triggering the rock to fall. As demonstrated by the 1872 Owens Valley earthquake, seismic shaking is a forceful rockfall trigger in Yosemite Valley (see vignette 7); earthquakes have triggered at least twenty rockfalls in the Valley since 1857. A recent example is the series of earthquakes centered near Mammoth Lakes that occurred between May 25 and 31, 1980. With magnitudes as great as 6.2, these earthquakes triggered at least nine rockfalls in Yosemite Valley. If you happen to be in Yosemite and feel an earthquake, get away from the cliffs!

Other rockfall triggers include water, ice, heat, vegetation, and animals. Of these, water is probably the most common. Virtually every winter storm triggers rockfalls in Yosemite Valley. Water that falls as rain or melts from snow enters fractures in rock. As the height of the water column in the fractures rises, it exerts an outward pressure that is often sufficient enough to dislodge rock masses that cling precariously to cliffs. Of lesser consequence is the role water plays in lubricating rock surfaces, reducing their frictional hold.

If you've ever forgotten about a bottle of wine or water chilling in your freezer and discovered the cracked or even shattered bottle in the morning, you know firsthand that water (wine too) expands when it freezes, and that this expansion exerts strong forces on the confining surfaces of the bottle. When water in a rock fracture freezes, it expands and forces the fracture open. Even if this expansion doesn't dislodge the rock from the cliff, it moves the rock slightly, changing its center of mass. When the ice melts, the rock remains in that new position. When water freezes in the now slightly larger fracture, it moves the rock a bit more. In this way, blocks of rock can be incrementally ratcheted from cliffs by the freeze-thaw process.

Because rain, snowmelt, and freezing temperatures are most common in the winter and spring, the number of rockfalls in Yosemite is generally greater during this time. The greatest number of rockfalls occur in the spring as snow and ice begin to melt. However, a significant number of rockfalls occur in other seasons as well.

What other mechanisms can trigger rockfalls? In certain cases biological forces can, mainly by a process called *root wedging*. As the name indicates, this process occurs when tree roots penetrate rock fractures in search of water. Roots expand as they grow, exerting pressure on the

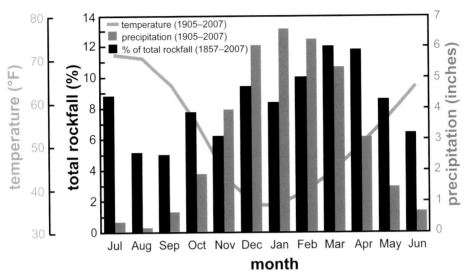

This graph shows that rockfall activity in Yosemite generally correlates with lower temperatures and greater precipitation amounts, but that rockfalls are common year-round.

sides of the fracture, which can ultimately trigger a rockfall. You have probably seen a similar process at work where tree roots lift and crack sidewalks or pavement. During strong winds, the trees can also act as levers, prying rocks out with their roots. Animals, including humans, scrambling over loose rocks can also cause them to fall. However, these incidents are relatively rare and certainly don't explain the 1996 Happy Isles rockfall.

It still isn't clear what triggered the 1996 Happy Isles rockfall. However, the Happy Isles event is not alone in this category; geologists can't reliably assign a trigger to more than half of the six hundred or so rockfalls documented in Yosemite since 1857. Without instruments documenting temperature, water pressure, and other variables within rockfall source areas at the time of failure, it is extremely difficult to determine a triggering mechanism with certainty. A number of large rockfalls, including the 1996 Happy Isles rockfall, occurred during the summer and in the absence of triggers such as storms or earthquakes. It may well be that thermal stresses (expansion and contraction) produced by warm summer temperatures and direct solar radiation are enough to break the physical bonds keeping large flakes of rock attached to cliff faces. Perhaps several mechanisms combine to trigger some rockfalls. Clearly, much work remains to be done to fully understand rockfalls in Yosemite.

Given their inherent unpredictability and destructive consequences, rockfalls present a real threat to human safety in Yosemite. As of 2009,

A rockfall tumbling down the northwest face of Half Dome to the floor of Tenaya Canyon near Mirror Lake on July 27, 2006. Like the 1996 Happy Isles rockfall, this event occurred on a warm, clear summer evening, without an obvious trigger. —Amanda Nolan photo

there have been fifteen documented deaths in Yosemite from rockfalls and related geologic events (such as debris flows), and many more people have been injured. Those numbers are actually fairly low compared to other natural hazards in the park, such as rivers and waterfalls, but a single rockfall event could tragically skew the numbers. For example, on November 16, 1980, a relatively small rockfall killed three hikers and injured at least seven others on the Yosemite Falls Trail. Rockfalls also pose challenges for those charged with managing Yosemite National Park. Some buildings on or near the talus slopes in Yosemite Valley have been damaged or destroyed by rockfalls. In the fall of 2008, the National Park Service permanently closed nearly three hundred buildings in the Curry Village area because of rockfall hazards, and park guidelines stipulate that no new building shall be placed on talus slopes. Part of the challenge for park managers is that in many areas of Yosemite Valley, rockfall hazard zones overlap with the flood zone of the Merced River, leaving few areas safe for development to accommodate large numbers of visitors. Ultimately, in a valley as deep, dynamic, and unpredictable as Yosemite, no place is absolutely safe from natural hazards.

As evidenced by the 1996 Happy Isles event, rockfalls are one of the most powerful geologic processes operating in Yosemite. Witnessing a rockfall can be an amazing experience; however, depending on your proximity it can also be very frightening, and in rare cases deadly. It is wise to respect this natural process and exercise caution when hiking or climbing in Yosemite Valley, or otherwise lingering beneath its walls.

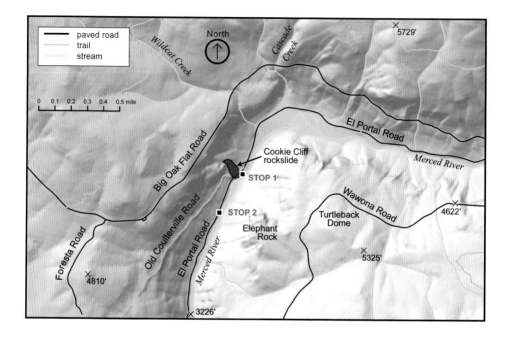

GETTING THERE

The Cookie Cliff rockslide, stop 1, is located between the west end of Yosemite Valley and El Portal. To get there from Yosemite Village, follow Northside Drive for 5.7 miles. Stay left (straight) at the junction with Big Oak Flat Road and follow California 140 west (El Portal Road), along the Merced River, for 2.3 miles down the gorge. Note the abrupt change in the river gradient between Yosemite Valley, where the river meanders gently over the valley floor, and the Merced Gorge, where the river is a nearly continuous series of rapids and cascades. Park in the large turnout on the left (east) side of the road that is adjacent to the river and near huge white boulders. To get there from El Portal, follow California 140 east (El Portal Road) for 5.8 miles and park in the large turnout (described above) on the right (east) side of the road. Stop 2 is a turnout on the east side of the road 0.3 mile southwest of stop 1.

6

That's the Way the Cookie Crumbles

THE 1982 COOKIE CLIFF ROCKSLIDE

Rockfalls and rockslides are common in and around Yosemite Valley—so common that virtually all of the major cliffs have experienced at least one rockfall since 1850, when people began keeping written records of these events. Many cliffs have experienced dozens of rockfalls in this time. And for those cliffs that haven't had rockfalls in historic time, large piles of rock debris (talus) beneath them attest to the great volumes of rock that fell prior to 1850. Most of the cliffs in Yosemite Valley still present rockfall hazards, but Cookie Cliff is an area where one can view the results of a large rockslide relatively safely. Note that we said *relatively* safely. Although rockfalls and rockslides are not as frequent at Cookie Cliff as they are in many other areas around Yosemite Valley, no cliff is absolutely free of rockfall activity. Approach this area with caution, as you would the area beneath any cliff.

Standing at the turnout, a series of steep cliffs and ledges extend upward to the west (away from the Merced River) over 1,600 feet to the rim of the Merced Gorge. The cliffs, dubbed the Cookie Cliffs, are a popular climbing destination in winter and early spring, being southeast-facing and lower in elevation than Yosemite Valley, so you may see climbers scaling them. Directly in front of you are several giant, light-colored boulders, some as big as small houses. These boulders came tumbling down in one of the largest rockslides in the recorded history of Yosemite.

The winter of 1981–82 was exceptionally wet, and by April of 1982 the ground was saturated. On April 3, at 10:20 p.m., a huge section of rock on the northern part of Cookie Cliff let loose, crashing down the slope and into the Merced River. Fortunately, no one was in the area at the time, so there were no injuries or fatalities. However, rock debris covered over 500 feet of El Portal Road, including boulders weighing up to 20,000 tons, the equivalent of a cube of granite nearly 60 feet on a side. Sewage pipes under the roadbed were severed, and effluent flowed into the Merced River for several days. The telephone line was also destroyed along this stretch of road, but power lines were spared because the slide occurred between transmission line towers. El Portal Road was closed to visitor traffic for several weeks as the boulders burying the road were blasted away. The rockslide also buried a section of the historic Old

Coulterville Road, which was not cleared of debris. In all, some 300,000 tons of rock debris moved in the rockslide.

Subsequent investigation of the rockslide source area revealed that the massive blocks of granite that came sliding down had rested upon a sloping bedrock surface that was deeply weathered to grus (decomposed granite). It's likely that this grus became saturated during the wet weather, reducing the friction between the bedrock surface and the overlying blocks and allowing the blocks to slide downhill.

Most people use the words *rockfall* and *rockslide* interchangeably, but geologists recognize a distinct difference between the two. *Rockfall* implies that the rocks primarily fell, as in down a vertical cliff, whereas *rockslide* implies that the rocks slid down a slope. The mechanics of these two types of movement are quite different. Because the 1996 Happy Isles event (vignette 5) was characterized by a free-falling rock mass, it is called a *rockfall*. Because the 1982 Cookie Cliff event was characterized by blocks sliding down a slope, it is called a *rockslide*.

The source area and debris path of the Cookie Cliff rockslide as seen from the top of Elephant Rock on the opposite side of the Merced Gorge. Rock debris covered some 500 feet of road for several weeks after the event. Note the car and motor home on the road for scale.

Left: *The Cookie Cliff rockslide shortly after the event (present El Portal Road drawn in for perspective).* Right: *The same area in 2008. It is clear from these two images the huge amount of rock debris crews had to blast away to restore El Portal Road.*

It is worthwhile to spend a few minutes wandering around the boulders on the uphill side of the road, being careful of the occasional loose boulder. Some of the boulders are of impressive size, which is an indication of how flawless the granite is here. Rockslides occurring in densely jointed rock will produce small blocks delineated by the joint intersections (as in the small, blocky diorite talus in the aptly named Rockslides area west of El Capitan, for example). In contrast, rockslides occurring in massive rock with few, widely spaced joints can produce enormous boulders. The Cookie Cliff rockslide occurred in El Capitan Granite. The sheer size and steepness of its namesake indicates that the El Capitan Granite is a strong, relatively joint-free variety of granite. The curving surfaces on the large boulders are fractures (not preexisting joints) along which even larger boulders split apart as they bounced and rolled down the slope. The sound of them cleaving must have been tremendous. The curving fractures are more evidence of the massive nature of the granite; had it been heavily jointed, the talus would be small and blocky.

The freshly broken surfaces of these boulders display the mineralogic makeup of the El Capitan Granite nicely (see vignette 1 for a detailed description of this granite). The rock consists of beautiful greasy-looking gray quartz, small sprays of black biotite, and two feldspar minerals, orthoclase and plagioclase, both of which are white in these outcrops. In this part of the park, the El Capitan Granite also contains an abundance of dark blobs. Geologists originally interpreted these blobs as foreign rocks that the granite magma picked up on its travels through the crust, but they are now interpreted as basaltic magma that was injected into the granite magma body and stirred into it. Known as *enclaves*, dark blobs like these are common in granites around the world. Enclaves commonly weather out in relief on a cliff face, forming handholds and footholds that climbers call "chickenheads" (see vignette 2).

Enclaves exposed in an angular, 6-foot-long block of El Capitan Granite that was deposited during the 1982 Cookie Cliff rockslide.

The upper end of the rockslide dam that constricts the Merced River channel just upstream of Cookie Cliff is a knickpoint where the river changes gradient. Here it transitions from a low gradient upstream (this page) *to a high gradient downstream* (next page).

Impressive as the 1982 Cookie Cliff rockslide was, there is evidence that even larger rockslides occurred here in the past. Between Yosemite Valley and El Portal, the Merced River cascades down a series of rapids and small waterfalls. Just upstream of Cookie Cliff, however, is the one stretch where the Merced River runs slowly through a relatively flat section. Large boulders at the downstream end of this flat section were part of a prehistoric rockslide originating from just east of Cookie Cliff. Large boulders such as these are simply too big for the river to transport, even during floods, so they constrict it and are probably responsible for the river's gentle character upstream of the constriction. Rock debris dams are relatively common in steep river canyons such as this. There are many examples in Yosemite, the most notable being the huge rock avalanche deposit that dammed Tenaya Canyon to form Mirror Lake (see vignette 7). Although it has not yet dammed the Merced River, the Ferguson rockslide about 14 miles downstream of Cookie Cliff still has the potential to do so (see vignette 19).

The downstream end of the rockslide dam is clearly visible from the Cookie Cliff rockslide parking area; it is the bouldery cascade upriver of you. If you walk a short distance (about 100 feet) upstream from the parking area and boulder-hop toward the river (do *not* attempt this during high water!), you can see the upstream end of the rockslide dam. It forms the boundary between the flat section of river that stretches upstream to Cascade Creek and the frothing rapids where the river is constricted by boulders from the prehistoric rockslide. A prominent change in river gradient, such as the one caused by the rockslide, is called a *knickpoint*. This constriction exacerbated flooding during the 1997 flood (vignette 8), causing severe damage to El Portal Road in this location.

The downstream end of the rockslide dam, which formed long before the 1982 Cookie Cliff rockslide, as viewed from the turnout on the south side of El Portal Road. The boulders through which the river cascades slid down from Cookie Cliff and are too large for the river to transport.

The 1982 Cookie Cliff rockslide wasn't the only recent geologic event in this area. At stop 2, just downstream of Cookie Cliff and on the other side of the river, is Elephant Rock, so named because it resembles an elephant when viewed from the Big Oak Flat Road. In the center-right of Elephant Rock is an unmistakable light-colored rockfall scar. Virtually all of the cliffs in and around Yosemite Valley are covered with lichens and water stains, so overall they appear somewhat dark gray. When a rockfall or rockslide occurs, the darker rock falls away, exposing fresh, clean, lighter-colored granite underneath. With time, the dark color returns as lichens recolonize the new rock surface and water runs over it. Exactly how long this process takes depends on the location of the scar (which way it faces, its elevation, and so on), but it can take many decades, sometimes even a century or more, for the scars to heal. The prominent scar on Elephant Rock resulted from two rockfalls, one in December of 1970 and the other in March of 1971. Together, these two rockfalls produced over 70,000 tons of debris. The light-colored boulders in the river at stop 2 are from these events, whereas the darker, lichen-covered boulders are from earlier, prehistoric rockfalls. According to local legend, the 1971 rockfall sent boulders of such great size into the river that fish were splashed out of the river and left stranded and flopping on the road.

For over thirty years after the 1971 event, no rockfalls occurred at Elephant Rock, or at least none were reported. Then, on the afternoon of August 9, 2006, more rocks tumbled down, originating from the upper part of the scar created by the earlier events. Compared to the rockfalls in the early 1970s, the 2006 event was much smaller, but it was still impressive. Unlike the earlier falls, it occurred on a hot, sunny day in August with no recent precipitation—not the sort of conditions one normally associates with rockfalls (see vignette 5). Events like these remind geologists that we have a lot more to learn about the forces that trigger rockfalls and rockslides in Yosemite.

The light-colored rockfall scar (center top) *on Elephant Rock and the resulting boulders in the Merced River.*

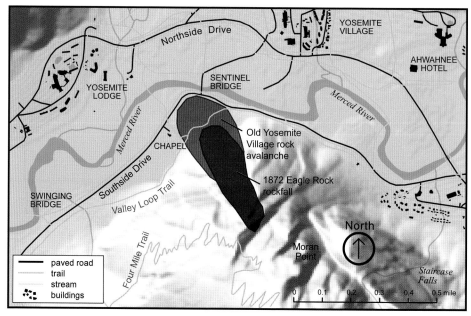

The approximate extent of the 1872 Eagle Rock rockfall (bright red) and the Old Yosemite Village rock avalanche deposit (light red), which originated earlier and from the same general location. (Modified from Wieczorek and others, 1999, and Wieczorek, 2002.)

GETTING THERE

There have been at least five rock avalanches—especially large and far-traveled rockfalls—in Yosemite. We discuss the deposits of three of them in this vignette. All three impressive locations can be visited in one day, and each location is a pleasant place to spend an hour or two. Though we give you directions to each locality, we recommend that you take advantage of the free shuttle bus. There are many day-use and trailhead parking areas in eastern Yosemite Valley from which you can walk or take the free shuttle bus to avoid traffic and parking problems.

The Old Yosemite Village rock avalanche is located near the Chapel in eastern Yosemite Valley. To get there from El Portal, follow California 140 east (El Portal Road) for 9.1 miles. Turn right, continuing on Southside Drive for 4.8 miles. Turn right into the parking lot for the Chapel. You can view the rock avalanche deposit from the Valley Loop Trail behind the Chapel, or wander out among the boulders immediately east of the parking lot, which are part of the deposit.

The Mirror Lake rock avalanche is located—no surprise here—at Mirror Lake. Unless you have a disability parking sticker, you cannot drive to Mirror Lake. The closest trailhead

7

The Walls Came Tumbling Down
EARTHQUAKES AND ROCK AVALANCHES IN YOSEMITE VALLEY

At 2:25 a.m. on March 26, 1872, one of the largest earthquakes in California's recorded history occurred on the Owens Valley Fault just east of the Sierra Nevada near the town of Lone Pine. The earthquake leveled most of the buildings in and around Lone Pine and killed twenty-three people. Although there were no seismometers at the time, it is estimated that the earthquake had a magnitude of about 7.5. Shock waves from the earthquake radiated out across the Sierra Nevada.

The earthquake triggered at least six rockfalls in Yosemite Valley. One of the largest was loosened from the west face of Liberty Cap near Nevada Fall. This rockfall had an approximate volume of 47,000 cubic yards, or over 100,000 tons; that's equivalent to a cube of rock about 100 feet on a side. It was notable not only for its large size, but also because it generated an air blast similar to, but smaller than, that caused by the

parking area is just east of Curry Village. To get there from El Portal, follow California 140 east (El Portal Road) for 9.1 miles. Turn right, continuing on Southside Drive for 6.1 miles, at which point you will pass the junction with Northside Drive. Continue heading straight, onto Happy Isles Loop Road, for another 0.5 mile to the trailhead parking area. From there, you can walk to Mirror Lake via the Valley Loop Trail, bike to it via a paved bike path, or take a free shuttle bus, which will drop you off about 1 mile from the lake, an easy walk along a paved road.

The El Capitan Meadow rock avalanche is located on the northern edge of El Capitan Meadow. To reach El Capitan from Yosemite Village, follow Northside Drive for 2.7 miles to El Capitan Crossover. Drive past this junction and park along the left side of the road adjacent to El Capitan Meadow. To get there from El Portal, follow California 140 east (El Portal Road) for 9.1 miles. Turn right, continuing Southside Drive for 2.4 miles. Turn left, taking the El Capitan Crossover to Northside Drive and park along the left side of the road adjacent to El Capitan Meadow. Cross the road and walk a short distance north toward El Capitan until you encounter the first large boulders marking the edge of the deposit.

1996 Happy Isles rockfall (vignette 5). Two years earlier, local proprietor Albert Snow had built a hotel at the base of Liberty Cap and Nevada Fall. Galen Clark, one of the early stewards of Yosemite, later described the 1872 event as it related to Snow's Hotel: "A large mass of rocks, which would weigh thousands of tons, fell from the west side of the 'Cap of Liberty' about a thousand feet above its base . . . When this great mass of rocks struck the earth, Mr. Snow says he was instantaneously thrown prostrate to the ground. The house . . . has moved two inches to the east." The light-colored scar from this rockfall is still visible on the west face of Liberty Cap.

On the morning of the earthquake, John Muir was sleeping in a small cabin near Black's Hotel, a long-gone inn that was located on the south side of Yosemite Valley near the present-day Swinging Bridge. The earthquake shook him out of bed, and he exuberantly bolted outside shouting, "A noble earthquake!" He later recalled his experience: "I feared that the sheer-fronted Sentinel Rock, towering above my cabin, would be shaken down . . . The Eagle Rock on the south wall, about half a mile up the Valley, gave way and I saw it falling in thousands of the great boulders I

The west face of Liberty Cap as seen from the John Muir Trail above Vernal Fall. The vertical, light-colored scar on the left side of the face marks where the rockfall detached during the 1872 Owens Valley earthquake.

had so long been studying . . . pouring to the Valley floor . . . After the ground began to calm I ran across the meadow to the river to see in what direction it was flowing and was glad to find that down the valley was still down." The earthquake and rockfall profoundly affected Muir's view of erosional processes, causing him to see earthquakes as the primary mechanism of rock debris (talus) formation in Yosemite Valley.

The rockfall witnessed by Muir originated from an area above the present location of the Chapel, from a point known as Eagle Rock. The rockfall was about 26,000 cubic yards in size, or roughly 59,000 tons, and was one of the larger rockfalls in Yosemite's recorded history. To put it in perspective, that's enough rock to fill about twenty-five four-bedroom houses. Despite its large size, it took geologists quite a long time to locate the exact source of the rockfall because of the ever-changing place-names in Yosemite. You can get a good view of the debris produced during the Eagle Rock rockfall by walking a short distance on the Valley Loop Trail behind and east of the Chapel. As you walk toward Curry Village, many of the large boulders to your right (south) are from this rockfall. All of the large trees growing among these boulders germinated after the 1872 event.

If you detour north (left) off the trail, toward Northside Drive and the Merced River, you are walking among boulders from an earlier and much larger event. These boulders are part of what is known as a *rock avalanche*, a special class of extra-large rockfalls. As discussed in vignette 5, most debris from Yosemite rockfalls piles up at the bases of cliffs, forming wedge-shaped deposits called *talus slopes* or *talus piles*. Debris from rock avalanches, however, extends out much farther from the bases of cliffs, onto the floor of the Valley.

The rock avalanche deposit near the Chapel has been named the Old Yosemite Village deposit because it extends from the south valley wall all the way to the south buttress of Sentinel Bridge, which was once the site of Yosemite Village (the village was eventually moved to the north side of the river to be in the sun and beyond the threat of floodwaters). Southside Drive is squeezed between the edge of the rock avalanche deposit and the river. The deposit has a volume of about 1.1 million cubic yards (one thousand four-bedroom houses, a small town's worth), or roughly 3.3 million tons. At present, all we know about the history of this rock avalanche is that it occurred after the Tioga-age glacier retreated from Yosemite Valley about 15,000 years ago and before events in Yosemite were first written down in 1851. If it had occurred before the glaciation, the ice would have transported the rock debris to the glacier's terminus, many miles down the valley.

The largest rock avalanche deposit in Yosemite Valley is located in Tenaya Canyon. At some prehistoric, postglacial time, a gargantuan body of rock on the north wall of the canyon, just east of Washington Column and probably of roughly the same size, collapsed into Tenaya Canyon.

One of the larger boulders in the Mirror Lake rock avalanche deposit is located along the Valley Loop Trail west of Mirror Lake.

Avalanche debris swept across the canyon and piled up against the far (south) wall, forming a deposit some 2,500 feet long, 3,000 feet wide, and 100 feet deep. The total volume of material is estimated to be about 15 million cubic yards (over thirteen thousand four-bedroom houses, a good-sized suburb), or roughly 34 million tons. The debris dammed Tenaya Creek, forming a large lake that once extended over 1 mile upstream of the dam. This lake has mostly been filled in with sediment carried by Tenaya Creek, and Mirror Lake is what remains.

Debris from the Mirror Lake rock avalanche is obvious along the paved road leading to Mirror Lake. At the junction of the Valley Loop Trail and the road that leads to Mirror Lake from the stables, you can start to see large boulders off to the north (left). Soon the road climbs up the dam created by the avalanche. As you approach Mirror Lake's outlet, just before the road ends in a small parking area, look south (right) across the creek. Tenaya Creek has exposed the interior of the rock avalanche deposit here, and it looks like what you'd expect a rockfall deposit to look like: jumbled and sheared angular rocks in a matrix of dust and small rock fragments that formed as falling rocks were pulverized. These rocks were probably emplaced in a matter of seconds. The violence of such an event is almost unimaginable.

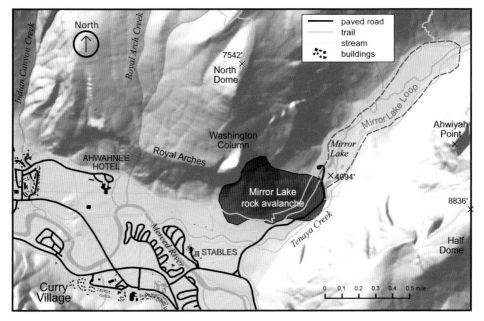

Debris from the Mirror Lake rock avalanche (red) filled a section of Tenaya Canyon to a depth of over 100 feet, damming Tenaya Creek and creating a large lake (dashed blue line). Mirror Lake is the filled-in remnant of this lake.

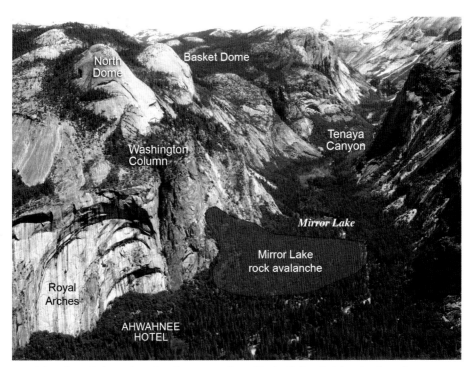

The Mirror Lake rock avalanche as seen from Glacier Point. Originating from just east of Washington Column, rock debris (red) of the avalanche fanned out across Tenaya Canyon, damming Tenaya Creek to form a large lake.

The Mirror Lake rock avalanche deposit—shattered, angular fragments in a matrix of pulverized rock—exposed in a cutbank along Tenaya Creek just downstream of Mirror Lake. A rock avalanche deposit can be distinguished from glacial till by the angularity of its rock fragments and, in most cases, the homogeneity of rock type. Till often contains rounded fragments of many different rock types.

The final rock avalanche deposit we discuss is perhaps the most interesting, in large part because of its spectacular setting beneath El Capitan. François Matthes, an eminent Yosemite geologist of the early 1900s, described the El Capitan Meadow rock avalanche deposit like this:

> The most remarkable body of earthquake débris is that which lies in front of El Capitan—not the talus of blocks that slopes steeply from the cliff to the valley floor, but the much vaster hummocky mass, partly obscured by a growth of trees and brush . . . that sprawls nearly half a mile out into the valley . . . There can be no doubt that it is the product in the main of one colossal avalanche that came down from the whole height of the cliff face—probably the most spectacular rock avalanche that has fallen in the Yosemite Valley since the glacial epoch . . . the quantity of débris that fell in this stupendous earthquake avalanche is so great . . . that its removal doubtless altered appreciably the contour and appearance of El Capitan.

The deposit extends nearly 2,200 feet from the base of El Capitan and has an average width of about 1,400 feet. It is at least 60 feet thick in places. From the leading edge of the talus slope at the base of El Capitan, the avalanche deposit slopes upward toward El Capitan Meadow, rising

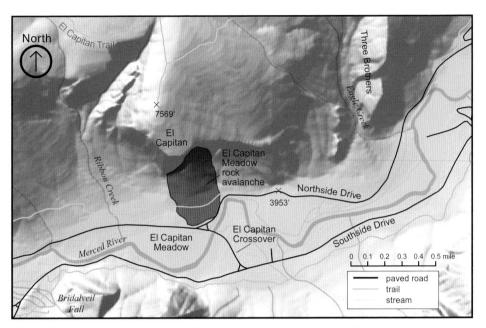

The approximate extent of the El Capitan Meadow rock avalanche (red).

nearly 20 feet to the crest of the deposit. The deposit has a volume of about 3.75 million cubic yards, or about 8 million tons. This is considerably smaller than the Mirror Lake rock avalanche deposit, but still huge— enough to fill three medium-sized college football stadiums. As with the other rock avalanche deposits, the El Capitan Meadow deposit is very coarse, with large, angular boulders up to about 3,500 cubic yards in size, or over 8,000 tons. Most of these large boulders are positioned along the edge of the deposit farthest from El Capitan.

Your second author has used cosmogenic exposure dating to determine when the El Capitan Meadow rock avalanche occurred. The technique, discussed in more detail in vignette 9, utilizes the fact that rocks exposed at Earth's surface accumulate unique isotopes with time (an isotope is a variant of an element with a different number of neutrons in the nucleus of the atom). By measuring the amount of these isotopes (beryllium-10, in this case) that occur in a rock, it is possible to determine roughly how long a rock has been exposed on the surface. Rock avalanche deposits are excellent candidates for this type of dating, for they mostly consist of rocks that were instantaneously exposed during the rock avalanche.

Beryllium-10 exposure dating of the El Capitan Meadow rock avalanche deposit indicates that the event occurred about 3,600 years ago. This age coincides nicely with age estimates for an earthquake related to an

The largest boulder on the surface of the El Capitan Meadow rock avalanche weighs some 8,000 tons.

ancient rupture of the Owens Valley Fault, the same fault that produced the 1872 earthquake. Dr. Jeffrey Lee of Central Washington University and his colleagues dug trenches across the fault and studied the subsurface sedimentary layers. By dating sediment layers that were offset by the fault, they were able to determine that a rupture preceding that in 1872 occurred between 3,300 and 3,800 years ago. These dates bracket the timing of the El Capitan Meadow rock avalanche as determined by cosmogenic exposure dating, suggesting a link between the two events. The dramatic events of March 26, 1872, clearly demonstrated that Owens Valley earthquakes are capable of shaking loose large rockfalls in Yosemite Valley. It appears likely that the same fault that triggered the dramatic Eagle Rock rockfall, which affected John Muir so profoundly, also triggered the massive El Capitan Meadow rock avalanche.

Other rock avalanches in Yosemite Valley have not yet been dated, but it would be interesting to know if the same earthquake that triggered the El Capitan Meadow rock avalanche also triggered others. The answer has important implications for the likelihood of such events occurring in the future, which would be catastrophic indeed.

_____8

How Water Sculpts Yosemite

THE FLOOD OF 1997

The role that moving ice played in creating Yosemite's landscape is well-known and hard to miss. Less appreciated, perhaps, is the role that water plays. Whether water or ice does more work on the landscape is the subject of ongoing debate among geologists, but on New Year's Day of 1997 there was no doubt that water was rearranging Yosemite Valley in profound ways. In this vignette we examine that flood and the effect it had on the landscape, roads, structures, and visitor usage of Yosemite National Park.

The groundwork for the flood was laid during a series of storms in November and December of the previous year. Calendar year 1996 was a wet year in California. The weather station in Yosemite Valley recorded 62.7 inches of precipitation, two-thirds more than in an average year. Nearly 22 inches fell in the Valley in November and December, more than 12 inches over the long-term average for those months. The main air route from the east to Bay Area airports flies over Tioga Pass and Tuolumne Meadows, and your first author observed the area twice in December of 1996 en route to San Francisco. In mid-December the high country around Tuolumne Meadows was covered by several feet of snow, but Tioga Road and other structures were still evident. By the end of the month—a mere two weeks later—several more feet of snow had fallen and Tioga Road was unrecognizable from the air. At the end of December about 100 inches of snow covered Tuolumne Meadows, and the snowpack in the mountains was about 180 percent of normal for that time of year.

A snow sensor at Tuolumne Meadows recorded the steady accumulation of snow by measuring snow water content. Snow water content is the depth of water that would be produced if a given amount of snow melted. This is easier for an automatic sensor to measure than snow depth, because all it has to do is weigh the snow, and it's also a more important number to know than snow depth, because it is a measure of the actual amount of water on the landscape. The relationship between snow depth and water content depends largely on the age of the snow and what it was like when it first fell. While 10 inches of light, fluffy, fresh snow can melt down to just 1 inch of water, it might take only 3 or 4 inches of older snow to make 1 inch of water.

111

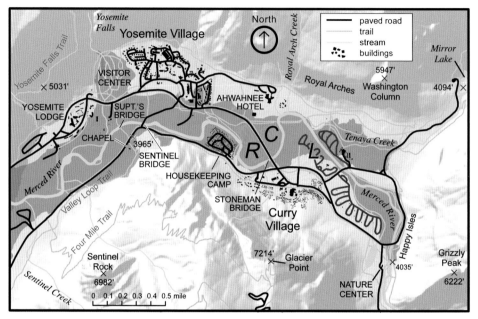

The extent of the January 1997 flooding (blue overlay) in eastern Yosemite Valley. R and C mark the locations of the former Upper and Lower River campgrounds. Part of Lower Pines Campground, marked L, was abandoned as well. "Supt.'s Bridge" is Superintendent's Bridge.

GETTING THERE

A walking tour of an hour or two in the area of the Chapel and Sentinel Bridge provides an excellent introduction to the effects of the 1997 flood of the Merced River. The parking closest to Superintendent's Bridge, where we begin our tour, is at the Chapel. To get there from El Portal, follow California 140 east (El Portal Road) for 9.1 miles. Turn right, continuing on Southside Drive for 4.8 miles. Turn right into the parking lot for the Chapel. Parking is also available at the Yosemite Falls viewpoint on the north side of the Merced River at Sentinel Bridge and day-use parking areas near Yosemite Village, from which you can walk or take the free shuttle bus to avoid traffic and parking problems. Excellent views of massive boulders that have moved during major floods can be seen along California 140 at and just upstream of El Portal.

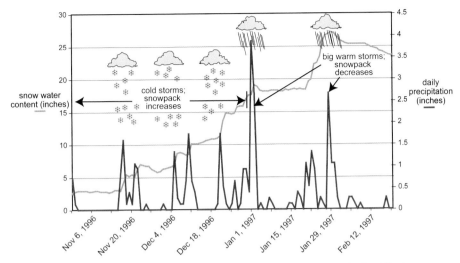

Snow water content (blue) and daily precipitation water content (red) at Tuolumne Meadows during the storms in the winter of 1996–97. The spikes in the graph clearly show that several storms hit the high country in November and December, each dumping an inch or two of precipitation and contributing roughly that amount to the snow water content, which climbed steadily. These storms were cold and deposited snow at Tuolumne Meadows. The New Year's Eve and New Year's Day storms were warm and it rained at that elevation, melting some of the snowpack.

Weather forecasters had predicted that the end of the year was going to be wet and wild because of the Pineapple Express, a weather pattern in which a branch of the jet stream directs warm, moist air from the central Pacific Ocean to the West Coast of North America. Typically, this pattern results in a series of warm storms spaced a day or two apart. The November and December storms were cold and milder than expected, but the storms that struck on New Year's Eve and New Year's Day were ferocious, dumping about 15 inches of precipitation on the western slope of the central Sierra Nevada and considerably more in the northern part of the state.

The towns of central California and western Nevada were pummeled by intense rain and high winds. The Russian River north of San Francisco crested 13 feet above flood stage (the level at which rivers spill over their banks and inundate their floodplains) and water was flowing more than 3.5 feet deep on aptly named Lake Street in downtown Reno. Several major highways were closed, and thousands of people were evacuated to high ground.

The New Year's storm was a "rain-on-snow" event, meaning warm, wet air from the Pineapple Express dropped rain on slopes that were covered in deep snow. Nearly 4 inches of water fell in Tuolumne Meadows on January 1, but the snow water content increased only a small amount and then decreased a bit on January 2. This reveals that precipitation at that elevation (8,600 feet) fell as rain, and that at least some of the snow-pack melted. So not only did a large amount of rain fall, but more water was produced by the melting of snow. Making matters worse, the snow and ground were both saturated. In effect, the snow shed the rain like a parking lot rather than a forest floor, which would absorb it and slow it down. A huge load of water was funneled by tributaries into Tenaya Creek and the Merced River and raced downhill toward Yosemite Valley, filling it nearly wall to wall.

The volume of water flowing down the Valley was truly colossal and capable of immense damage. To illustrate this, we borrow an analogy from Jeff Mount's book *California Rivers and Streams* (a must-read for anyone interested in surface water). Peak flow at Pohono Bridge at the western end of the Valley was estimated at about 24,600 cubic feet per second. A cubic foot of water weighs about 63 pounds, so that means about 1.55 million pounds of water was flowing past the bridge *each second*. This is the equivalent of about 337 Ford Explorers—each second. Imagine the work—and devastation—that could be done on the valley floor and walls by such an armada. Another way to try to grasp the rate at which water flowed through the valley is to consider that the river could have filled a sixty-thousand-seat football stadium in about eighteen minutes.

Flooding was extensive in the occupied eastern end of the valley. Flood-waters covered North Pines and Lower Pines campgrounds and completely overwhelmed Upper and Lower River campgrounds. Housekeeping Camp, the area around the Yosemite Village store, and part of Yosemite Lodge were also flooded. The meadows were awash on both sides of the Val-ley. The Ahwahnee Hotel, employee housing and administrative areas in Yosemite Village, and Curry Village were largely spared, being on slightly higher ground. An 8-mile stretch of El Portal Road (California 140) was severely damaged or washed out. Hundreds of rooms at Yosemite Lodge were flooded with 5 feet of water, as were many National Park Service administrative and residential buildings around Yosemite Village. Towers supporting high-voltage electrical lines were extremely damaged. Because of the damage to infrastructure, the valley was closed to visitors for more than two months. The damage was estimated at about $178 million.

We begin our walking tour at Superintendent's Bridge, the footbridge just downstream of Sentinel Bridge that provides a route between the Chapel and Yosemite Village. A sign near the northern end of Superinten-dent's Bridge shows the level at which the river peaked during the five major floods of the 1900s. The 1997 flood was the largest, having peaked 5 feet above the bridge. It is sobering to stand on the bridge and imagine swiftly moving, debris-choked water flowing that high.

Cross the bridge and walk south to the Chapel. Equally sobering is the sign here showing the high-water mark of January 2, 1997. Water was almost 10 feet deep here; floodwaters reached the door handle, and the

Debris covering Superintendent's Bridge on January 7, 1997. The bridge was significantly damaged. —Steve Thompson photo, courtesy of Yosemite Research Library

Chapel was extensively damaged. The Chapel was first used in 1879, when it was located to the west, near the base of the Four Mile Trail to Glacier Point. It was moved to its present site (near other buildings in what was once Old Yosemite Village) in 1901. In 1965–66 the foundation was raised 3 feet to better protect the structure from flooding, but this wasn't enough to prevent severe damage in the flood of 1997. (Before you leave this area, you might want to visit the nearby site of the 1872 rock avalanche discussed in vignette 7.)

Walk east from the Chapel along the trail on the south side of the road (don't walk on the traffic-laden road!), past LeConte Memorial Lodge and Housekeeping Camp. The steep talus apron of the south wall cliffs is on your right. At the four-way intersection north of Curry Village, about 1 mile from the Chapel, turn left, back onto the river floodplain, and cross Stoneman Bridge. After crossing the bridge, you are standing in the floodplain of a large meandering bend of the Merced. Upper and Lower River campgrounds were located on this floodplain, which proved to be a dangerous place. When the river topped its banks it flowed with great force across this flat area, inundating and obliterating the closed campgrounds. One might ask why Lower Pines Campground, located inside the upstream meander to the east, was not as severely damaged. One factor is that Tenaya Creek joins the Merced partway around that bend, so the volume of water downstream of the confluence was much greater

The entrance to Lower River Campground on January 2, during the flood. The water was about 3 feet deep. —Steve Thompson photo, courtesy of Yosemite Research Library

This campsite wasn't a very comfortable place to pitch a tent after floodwaters tore up asphalt and deposited a load of cobbles on it. The river ripped up roads, smashed buildings, and carried away picnic tables in the former Upper and Lower River campgrounds.
—Courtesy of Yosemite Research Library

than above. Even though Lower Pines Campground was inundated, it experienced a less powerful flow of water than the Upper and Lower River campgrounds.

Damage was extensive outside the Valley as well. At least thirty-six sections of California 140 between the Cookie Cliffs rockslide and El Portal, a distance of only about 6 miles, were heavily damaged. In many areas this road damage exposed important pipelines, including a 12-inch sewer line that served the Valley. At the rockslide itself (vignette 6), a 700-foot section of road was obliterated, the same piece of road that was obliterated by the 1982 rockslide. Rockslides and flooding are clearly related here, because the natural dam produced by repeated rockslides in this area constricts the river and causes it to flow, at high water, along a secondary channel that is aimed squarely at this section of road.

The banks of the Merced River near the town of El Portal consist of huge boulders that represent most of the rock types found upstream. Glaciers can deposit similarly jumbled rock types in a moraine, but the roundness of these boulders, their size sorting, and their confinement to the banks of the river show that they were carried here by floodwaters, having been moved hundreds to thousands of feet. It is difficult to conceive of the forces needed to move such boulders, but El Portal residents watching the raging floodwaters in 1997 described hearing boulders smashing into other boulders underwater and even seeing sparks produced by these impacts.

This section of California 140 at the Cookie Cliffs rockslide was destroyed in the 1997 flood. Although this is a bad place for the road, there is nowhere else to put it.
—Courtesy of the Yosemite Research Library

Nearly all of these boulders lining the Merced River near El Portal, many of them 5 or more feet in diameter, were carried here during floods. These are part of the river's bed load, the material moved along the riverbed by rushing water. The 1997 flood significantly rearranged these boulders.

The flood of 1997 is commonly described as a 90-year event. What does this mean? A common misinterpretation is that a flood of the same magnitude should not be expected for about 90 years, but that isn't correct: a 90-year flood is one that can be expected to occur *on average* every 90 years. Take the *100-year flood*, a term often mentioned in media accounts of floods. If a flood of a certain magnitude is expected to occur on average every 100 years, then in any given year there is a 1 percent chance it will happen. That means 1 percent this year, 1 percent next year, and so on. If such a flood were to occur this year, there is still a 1 percent chance that it could happen next year. In any 50-year stretch, there is about a 40 percent chance that such a flood will occur; in a 70-year stretch, the odds are about even.

Flood frequency for a given river is estimated from historical data. In Yosemite Valley, the Happy Isles gaging station on the Merced River has been in operation since 1915. Among the bushels of data available for Happy Isles are measurements of peak discharge for each year since 1916. This measurement represents the highest measured flow for the entire year. Scientists use peak discharge numbers to determine flood frequency on undammed rivers. (For information on how rivers behave and how their flows are measured, see the introduction.)

How do we use peak discharge data to estimate how often floods occur? This is generally done by estimating a statistic known as *recurrence interval*, which in this case is a measure of flood likelihood. The general scheme is pretty simple even though it is based on advanced statistical methods: take a data set, such as that for peak discharge at Happy Isles, and rank the results from highest to lowest, giving the highest flow a rank of 1, the next highest a rank of 2, and so on. Recurrence interval is then estimated with the formula $\frac{n+1}{r}$, where n is the number of years in the data set and r is the rank.

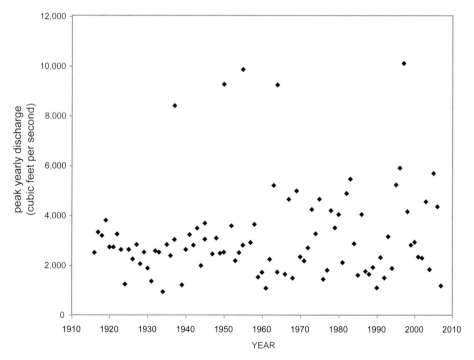

Peak discharge at Happy Isles from 1916 to 2007. Several things are apparent from this graph. First, peak discharge varies a lot from year to year. Second, the years 1937, 1950, 1955, 1964, and 1997 had much higher peaks than the others, with 1997 being the highest. These are the five floods whose high-water marks are shown on the sign near the north end of Superintendent's Bridge. Third, eyeballing the graph shows that average peak discharge seems to be about 3,000 cubic feet per second. Finally, the variability in discharge has increased since about 1960. This pattern of more variable discharge (and, by inference, more variable weather) in the latter half of the twentieth century is also evident in data from other rivers in California. Some years are definitely a lot wetter than others, with the Merced carrying seven to eight times more water in the wettest years than in the driest ones. (Data from the National Water Information System.)

For example, in 2007 the estimated recurrence interval for the 1997 flood was 93 years. If a flood of that magnitude had occurred in 2007, it might have been called a "93-year flood" in the media. The next-largest flood has an estimated recurrence interval of half that, or 46.5 years, and the third-largest, 31 years. If 10 more years pass without the peak discharge of the 1997 flood being exceeded, then the estimated recurrence interval for a flood of that size will have increased to 102 years. It is evident that this sort of forecasting is an imprecise science, with the estimated recurrence interval of the largest flood in the data set always exceeding by 1 year the time that the river has been observed.

DATE	PEAK VALUE (cfs)	RANK	RECURRENCE INTERVAL
January 2, 1997	10,100	1	93.0
December 23, 1955	9,860	2	46.5
November 18, 1950	9,260	3	31.0
December 23, 1964	9,240	4	23.3
December 11, 1937	8,400	5	18.6
May 16, 1996	5,900	6	15.5
May 16, 2005	5,680	7	13.3
May 29, 1983	5,450	8	11.6
July 9, 1995	5,220	9	10.3
February 1, 1963	5,200	10	9.3

Ranking of the ten largest peak discharges since the Happy Isles gaging station began operating, and their recurrence intervals.

Studies of soils, tree growth, and archeological sites along the Merced and several other Sierra Nevada rivers indicate that floods comparable to the 1997 flood are fairly common, with recurrence intervals of decades to a few hundred years. However, significantly larger floods seem to be uncommon, as archeological sites and other evidence of stability over time periods of hundreds to thousands of years are common at elevations just above those reached by floodwaters in 1997. This suggests that the 1997 flood was about as large as any flood in the last several thousand years. Scientists studying the effects of global warming in California project that warming in the next century will cause the snow line (the elevation separating snow from rain) to rise a few thousand feet. If this happens, the area in Yosemite receiving precipitation as snow will shrink dramatically, and most of the area now receiving snow will receive rain instead. Considering that a warm rainstorm triggered the 1997 flood, there is concern that these types of floods may become more frequent.

9

A Natural Dam Across Yosemite Valley

THE EL CAPITAN MORAINE

As discussed in the introduction, Yosemite National Park was repeatedly blanketed with glacial ice during the last 2 million years. Climate records pieced together from evidence encased in ice and ocean sediment suggest that Yosemite was glaciated at least a dozen times during this period, and perhaps many more. However, because each advance of ice can wipe out evidence of previous glaciers, the geologic record of glaciations in Yosemite and the Sierra Nevada is much more limited. The on-the-ground evidence tells us there were at least three, and perhaps as many as six, glaciations preceding the most recent Tioga glaciation, which peaked about 20,000 years ago. Evidence of the Tioga-age glacier, which in Yosemite Valley had retreated by about 15,000 years ago, is obvious in the high-elevation parts of the park in the form of glacial polish, striations, erratics, and moraines. Evidence in Yosemite Valley is subtler. This vignette highlights the best evidence that there was a glacier in Yosemite Valley: its moraines.

A moraine is a ridge of debris left behind by a glacier. Some of this debris is material the glacier eroded by plucking and abrading the rock it flowed over, and some of it fell onto the glacier from adjacent ridges and peaks. Whatever the source of the debris, the glacier then transported it down the valley like a conveyor belt. When rivers carry debris, they neatly sort it according to the shear stress and velocity needed to transport particles of different sizes, leaving stratified deposits of particles that are grouped by size. In contrast, glaciers dump particles of all sizes in one big jumble, much as a dump truck would. As a result, moraines are usually an unsorted mess of silt, sand, gravel, cobbles, and boulders. This lack of sorting and layering is a key characteristic of moraines. Another characteristic is that moraines often contain large boulders of rock types not found locally, indicating that they traveled a long way in or on a glacier.

A terminal moraine is deposited at the end of a glacier, called its *terminus*, and lateral moraines are deposited along its sides. As glaciers retreat, sometimes they pause and deposit a recessional moraine. The El Capitan moraine is sometimes said to be a terminal moraine, but it is actually a recessional moraine that formed as the Tioga-age glacier paused briefly

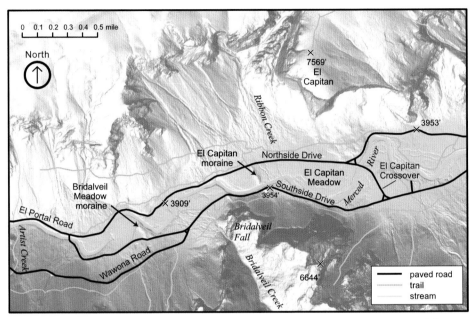

The El Capitan moraine is a recessional moraine of the Tioga-age glacier, and the Bridalveil Meadow moraine is generally thought to be the glacier's terminal moraine.

GETTING THERE

The El Capitan moraine is located in the western end of Yosemite Valley, just west of El Capitan Meadow. To get there from Yosemite Village, follow Northside Drive for 2.7 miles to El Capitan Crossover. Just past this junction the Valley opens up into El Capitan Meadow, with El Capitan soaring above to your right (north). At the west side of the meadow, the road enters the forest again and climbs a small hill. At the top of this hill, about 0.5 mile from the crossover, look for a small sign on your right labeled "V9." Park in a large gravel turnout on the right immediately after this sign. After checking for traffic, carefully cross the road and walk south onto the crest of the hill, which is the El Capitan moraine. To get there from El Portal, follow California 140 east (El Portal Road) for 9.1 miles. Turn right, continuing on Southside Drive for 2.4 miles. Turn left, taking the El Capitan Crossover to Northside Drive heading west and follow the directions above.

in its retreat from the Valley. This glacier's terminal moraine is probably the larger but less obvious Bridalveil Meadow moraine, 0.75 mile to the west, just down the valley from Bridalveil Fall.

In places, as much as 2,000 feet of sediment lies atop the bedrock in Yosemite Valley, filling a deep bedrock depression (see the introduction). Most of the scouring of this deep depression probably occurred during earlier glaciations, perhaps about 1 million years ago, with the 2,000 feet of sediment being deposited by rivers, streams, and rockfalls afterward. The advancing Tioga-age glacier, much smaller than earlier glaciers, excavated only a slight depression in the sediment fill east of the El Capitan moraine, coming nowhere close to cutting all the way down to bedrock. As the Tioga-age glacier retreated from Yosemite Valley, meltwater filled this basin, forming a shallow lake that François Matthes, a geologist studying Yosemite's landscape in the early twentieth century, christened Lake Yosemite. Matthes and other geologists used to think this was a wall-to-wall lake in the Valley, but our view is evolving. We now think it probably resembled a complex of interconnected ponds and braided streams, such as the terrain often seen in freshly deglaciated areas in Alaska. The El Capitan moraine acted as the dam behind which the shallow lake backed up, with the Merced River outlet flowing over a low spillway through the moraine near the south valley wall.

The El Capitan moraine was also a barrier to the down-valley transport of sediment. As the glacier retreated, the Merced River and its tributaries delivered large amounts of sediment into the lake basin, eventually filling it in and creating the relatively level valley floor of today. The extent of the sediment trapped behind the moraine can be easily seen from your position on the moraine. As you face south, toward the Merced River, look to your left (east) and right (west). The floodplain surface upstream of the moraine is visibly higher than it is downstream. Once the sediment reached the level of the lake spillway, it no longer accumulated behind the moraine and was transported downstream.

Take a walk down the length of the El Capitan moraine, which is easy owing to a number of fires that have burned most of the brush and dead wood on the moraine, most recently (before this writing) in 2004. Overall, the moraine is fairly long and narrow, with a smoothly convex, parabolic surface. Except for a gap carved by the river, the moraine extends all the way to the south wall of the valley. This cross-valley orientation is one sure sign that you stand on a moraine, because rivers generally don't deposit large ridges of sediment perpendicular to their channels.

As you walk down the moraine, take note of the variety of rock sizes and types exposed on the surface. After about 100 feet, look for boulders of Cathedral Peak Granite, which have conspicuous rectangular crystals of orthoclase, some as large as 4 inches long (see vignette 1). The pink to white crystals are somewhat more resistant to weathering than the surrounding smaller minerals, so they often jut out from the boulder

The view southward down the length of the El Capitan moraine, showing its broad, convex surface.

surfaces. The presence of many different rock types, especially the Cathedral Peak Granite, is another sure sign that you stand on a moraine and not, for instance, a local rockfall deposit. The nearest bedrock outcrop of Cathedral Peak Granite is some 12 miles to the east. These boulders were either plucked from the bedrock beneath the glacier or fell from peaks and ridges onto the glacier to be transported and dumped in Yosemite Valley as the Tioga-age glacier retreated. It is unlikely that the Merced River transported them here; due to the river's low gradient through the Valley, it doesn't have sufficient power to move boulders this size, even during floods.

As you walk south, you will eventually encounter a very large cubic boulder, some 8 feet high, sitting on the crest of the moraine. Being the largest on the moraine, this boulder was a good target for cosmogenic exposure dating, a relatively new dating technique developed in the 1990s. New dating methods developed over the past several decades have allowed geologists to determine the ages of many types of rocks, but geomorphologists, geologists who specialize in Earth's surface processes, were left out of this party. Rather than wanting to know the ages of the rocks themselves, geomorphologists are often more interested in knowing how long certain rocks have been in a certain place; for example, how long this large boulder has been sitting on this moraine. The

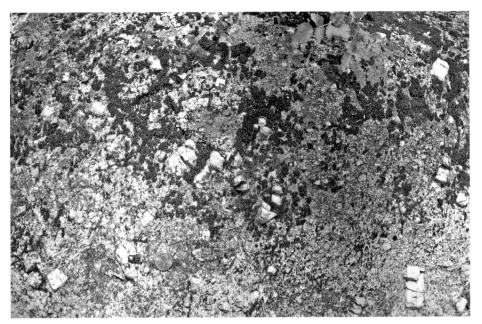

A boulder of knobby, lichen-encrusted Cathedral Peak Granite on the surface of the El Capitan moraine. Lichen doesn't grow as easily on the feldspar crystals as it does on the surrounding rock. Penny for scale.

granitic boulders in the El Capitan moraine formed roughly 102 to 88 million years ago, but they have been part of the moraine for only a very small fraction of that time.

Geomorphologists finally found a reliable dating tool with cosmogenic exposure dating, a technique so weird and wonderful it merits a brief discussion. Some exceedingly rare radioactive isotopes on Earth, such as beryllium-10, form when cosmic rays strike certain elements, such as oxygen, on or near Earth's surface. Isotopes are variants of elements that differ in the number of neutrons they contain. Cosmic rays are highly energetic particles that come from exotic extraterrestrial sources, such as neutron stars and supernovae. Rocks on Earth's surface are bombarded by cosmic rays, and beryllium-10 accumulates within the oxygen in surface minerals such as quartz at essentially constant, but exceedingly slow, rates. The longer a rock is exposed at the surface, the more beryllium-10 it contains; however, even in rocks that have been exposed for a long time, beryllium-10 accounts for only about 1 atom out of every trillion or quadrillion atoms. Finding those scant atoms in a sample is akin to finding one particular sand grain in a swimming pool full of sand. Only a handful of labs around the world can make these measurements.

The buildup of beryllium-10 in a rock is a bit like getting a suntan; if you assumed that a person started off with a particular shade of skin and you knew the rate of tanning, you could "measure" their tan and estimate how long this person had been exposed to the sun. The analogy can be taken even further: if you get sunburned, your skin will peel off, complicating estimates of sun exposure. Likewise, hot forest fires cause slabs of rock containing accumulated beryllium-10 to spall off of boulders, complicating exposure dating. For that reason, and also because moraines tend to change shape with time, from sharp-crested to rounded, the tops of the largest boulders on a moraine crest are the best candidates for dating. Your second author used the cosmogenic beryllium-10 technique to date three boulders on the El Capitan moraine, including the largest boulder before you. The results suggest that the moraine was deposited about 19,000 years ago.

The river's edge, a short scramble down the eastern slope of the moraine, is a good place to view the southward extension of the moraine across the Merced River. The river and Southside Drive pass through the moraine where the river has cut it, and the moraine's convex cross-sectional shape is visible. A small rapid occurs where the river crosses the moraine, and the river upstream is relatively flat. Over the 19,000 years

Large boulders on the crest of the El Capitan moraine. Cosmogenic beryllium-10 exposure dating indicates that these boulders and the moraine are about 19,000 years old. The largest boulder is about 8 feet tall.

View southward across the Merced River, showing the convex cross section of the El Capitan moraine. Southside Drive passes through the gap that the river has cut through the moraine.

or so since the Tioga-age glacier receded from this area, the river has cut this narrow channel, washing boulders, cobbles, and finer sediment out of it. Only the very largest boulders remain, those that were too large for the river to transport even in flood stage. These boulders, called *lag boulders* because they are what's left behind after the smaller stuff has been carried away, partially block the channel and form the small rapid.

This channel is one of the narrowest spots in the river anywhere in Yosemite Valley and was therefore the site of a former bridge. Prior to 1878 there were just two bridges across the river in Yosemite Valley, both in the eastern part of the Valley. The managers of the fledgling park decided to build a new bridge to foster tourism, constructing it across the El Capitan moraine, just above the rapids, in the fall of 1878. From your viewpoint along the river's edge, you can see the old dirt roadbed running along the eastern edge of the moraine; this roadbed once led across the bridge.

Even though the river lowered its level by cutting through the moraine, the moraine still acted as a primary control of the river's level in Yosemite Valley. During winter floods and spring runoff, the river backed up against the moraine, flooding the Valley and leaving large areas of standing water.

This affected the ever-increasing human use of the Valley, closing roads and trails, washing out bridges, isolating hotels and cabins, and flooding the meadows, then used as pastureland for transport stock. Standing water and wet meadows supported abundant populations of insects, especially mosquitoes. And because the river was also the valley sewer at that time, the damming effect of the moraine contributed to sanitation problems.

Those in charge of managing Yosemite at that time blasted the rocks above the rapids at the new bridge to reduce yearly flooding. Galen Clark, one of those early stewards, described the blasting effort:

> When the El Capitan bridge was built in 1879 [sic] it was located across the narrow channel of the river between the two points of what remains of an old glacial . . . moraine. The river channel at this place was filled with large boulders, which greatly obstructed the free outflow of the flood waters in the spring, causing extensive overflow of the low meadow land above, and greatly interfering with travel, especially to Yosemite falls and Mirror lake. In order to remedy this matter the large boulders in the channel were

A small rapid has formed where the Merced River cuts through the El Capitan moraine. Before the river cut through the moraine, it acted as a dam, backing up a lake and trapping sediment behind it. The natural down-cutting of the river was artificially aided when the largest boulders were dynamited in 1878 to encourage drainage out of Yosemite Valley. View is looking in the upstream direction.

blasted and the fragments leveled down so as to give a free outflow of the floodwaters. This increased the force of the river current, which now commenced its greater eroding work on the riverbanks, and as the winding turns became more abrupt the destructive force annually increases.

Subsequent investigations of the El Capitan moraine suggest that the blasting lowered the river level by 3 to 5 feet, and ripple effects, such as the deepening of tributaries and lowering of groundwater levels, propagated far upstream of the moraine. A few people have even suggested that the National Park Service "restore" the moraine by adding boulders back to the area where the blasting occurred. Yet, despite the blasting the moraine still profoundly affects the hydrology of Yosemite Valley. The extent of flooding in eastern Yosemite Valley during the 1997 flood was so great largely due to water backing up behind this relict of the most recent glaciation.

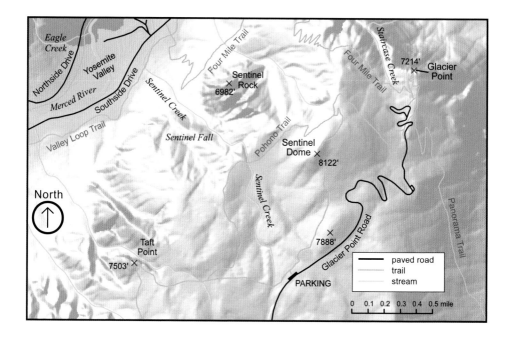

GETTING THERE

Taft Point lies a few miles southwest of Glacier Point along the south rim of the Valley. To get there from Yosemite Village, follow Northside Drive for 4.7 miles (from El Portal, follow California 140 east for 9.1 miles), then turn onto Southside Drive and follow it for 1 mile. Turn right onto California 41 south (Wawona Road), drive 9.2 miles, and turn left onto Glacier Point Road. The Taft Point/Sentinel Dome parking area lies about 13.4 miles down this road. This is a popular trailhead, and parking can be difficult to find in the high season; other options include parking near Glacier Point and making a longer hike that includes the hike discussed in vignette 11. From the Sentinel Dome/Taft Point parking area, hike about 80 yards north (downhill) across bare ground to the trail that goes left (west) to Taft Point (1.1 miles). This is an easy hike over gentle terrain, but at the point itself there are abundant unguarded cliff edges with 3,400-foot drops. Keep a close eye on children. The view can be enjoyed without venturing too close to the cliff edge.

10

Cracks in the Earth

THE FISSURES OF TAFT POINT

The cliffs of Yosemite are remarkable features; there are few places in the world where you can drive almost to the edge of a cliff that is more than 3,000 feet tall. Glacier Point is where most people view a cliff's edge in Yosemite because it's very accessible, but serious students of cliffs will also want to undertake the strenuous hikes required to view the abyss from the edge of El Capitan, Half Dome, Yosemite Falls, and other more remote vantage points. In between these hiking extremes lies Taft Point, which exhibits the "cliffiness" of the other sites and also offers a splendid view of central and western Yosemite Valley. Steep, seemingly bottomless clefts cut deeply into the cliffs, making it an outstanding place to study the control that steep joints exert on Yosemite's landscape.

Joints occur in all orientations. We discuss exfoliation joints, those that parallel the land's surface, in vignettes 11 and 12. In this vignette we discuss the steep regional joints that crisscross the landscape and control the orientations of features such as the face of Half Dome, the steps over which Nevada and Vernal falls drop (vignette 4), and the course of Yosemite Creek.

Steep joints are easy to spot on air photos because they appear as deep scratches or vegetation-filled gullies. They are also easy to pick out on digital elevation models, especially detailed images that are made by bouncing a laser beam off the landscape from a low-flying airplane. A quick glance at imagery of the area around Yosemite Valley shows prominent joints in a variety of orientations. The most prominent is the one that Yosemite Creek follows for a long distance. Tioga Road takes a big swing northward to get around this joint, now a 600-foot-deep valley. Steep joints also control the orientations of Eagle Creek, Indian Canyon, LeConte Gully, and El Capitan Gully, all tributaries entering Yosemite Valley.

To talk about joints we need to introduce a bit of nomenclature. The orientation of a planar geologic feature, such as a joint, is described by two quantities: strike and dip. Strike is the compass direction of a joint, and dip is how far the joint is inclined from the horizontal (horizontal equals 0 degrees; vertical equals 90 degrees). The joints we are discussing here are nearly vertical, so we can dispense with dip and only talk about strike. Strike is described by degrees clockwise from north, so a

Joints on a regional scale and in a variety of orientations dominate the landscape between Yosemite Valley and Tioga Road. For much of its length, Yosemite Creek follows a linear joint that can be traced for 12 miles. Note the large northward excursion that Tioga Road makes to avoid this valley.

joint oriented north-south has a strike of 0 degrees and one oriented northeast has a strike of 45 degrees.

The prominent vertical joint that controls the course of upper Yosemite Creek has a strike of about 30 degrees, which is a little north of northeast. A number of prominent joints west of Yosemite Creek parallel this joint, and the joint that controls the face of Half Dome is somewhat parallel to it, as is the joint that controls the southward jog the Merced River takes once it leaves the Valley, upstream of the town of El Portal. A number of prominent joints north of the Valley strike to the northwest at about 345 degrees. Many of the streams of the highland north of the Valley, including the main Tuolumne River, South Fork Tuolumne River, and Tamarack Creek, follow these joints.

South of the Valley the situation is similar. The highland south of Glacier and Taft points is transected by a grid of vegetation-filled joints. One set strikes 325 to 360 degrees (northwest to north), and another 40 to 60 degrees (northeast). This latter set is much more prevalent and nearly parallels the northwest face of Half Dome and the step over which Nevada Fall plunges. Sections of Glacier Point Road take advantage of these gashes in the landscape. The western two-thirds of the road runs eastward from Chinquapin before bending northward just before the Mono Meadow parking area. At that point it is following a north-striking joint that can be traced to Taft Point. The road climbs about 500 feet along this stretch before making a sharp jog to the east through a notch east of Ostrander Rocks. This notch leads to the next major north-trending joint, which the road follows through Pothole Meadows before swinging northeastward and passing the Sentinel Dome/Taft Point Trailhead on the way to Glacier Point.

Along the north side of the trail to Taft Point, about 100 yards west of the parking area, lies a prominent outcrop of brilliant white quartz. This massive quartz vein is similar to quartz veins that cut metamorphic rocks west and east of the park boundary (see vignettes 19 and 25, respectively). Veins outside the park contain ore deposits, including gold, but this vein is barren. Large quartz veins are rare within the granites of the park, although several occur in this area, including on Sentinel Dome.

The hike to Taft Point crosses four significant tree-filled gullies, all of which are north-striking and part of the set of joints striking northwest to north. The first (1) is just past the quartz vein, where the trail passes from open ground into dense trees. The little stream in the gully is Sentinel Creek, and in late spring it pours over Sentinel Fall. After emerging from

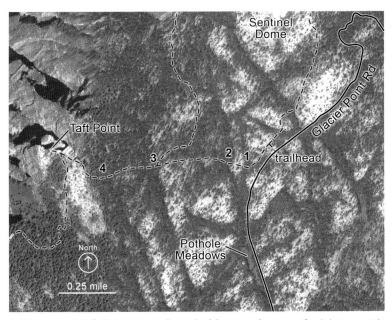

Joints around Taft Point are well marked by trees because the joints contain thicker soil than the surrounding granite. These joints are more obvious from the air than from the ground. Numerals 1 through 4 designate the tree-filled gullies referred to in the text.

that thicket, the trail crosses a few hundred feet of bare ground before heading back into trees at another joint (2). This joint and the previous one are branches of the joint that Glacier Point Road follows north from the notch by Ostrander Rocks. Following the trail west, the trees thin for about 300 yards before the trail comes to a trail junction with a sign (0.5 mile to The Fissures, 0.6 to Taft Point, and 5 to Dewey Point) and crosses another small stream in another tree-filled joint (3). This trail junction lies at the intersection of a prominent north-striking joint and a northwest-striking joint that defines one of the cliff faces at Taft Point.

After the junction the trail traverses several hundred yards of diverse evergreen trees growing on a narrow block of granite (4) that is defined on the east and west by two prominent, closely spaced north-striking joints. A last small climb brings you to open ground, with an easy descent on natural rock steps to the cliff edges. Hang on to the kids! The ground around here is badly trampled; please try to stay on existing trails.

This cleft, the scariest of The Fissures at Taft Point, is about 3 feet across. Two large boulders are wedged in it. In olden times climbers wedged rocks in cracks and tied slings around them to aid in climbing; such rocks were called chockstones. *These look like giant chockstones, and given the predilection for invoking Beelzebub in place-names, it's a wonder that this area wasn't called the Devils Chockstones (see vignette 18).*

Taft Point, a prow that juts out from the Valley's southern edge, is defined by a northeast-facing cliff and a west-facing cliff. Were it not recessed into a large indentation in the southern wall, it would be quite a landmark. The trail stays to the left of The Fissures, unmistakable deep clefts excavated by erosion along joints that extend deep into Taft Point's northeast-facing cliff. There are three prominent clefts cut into this face. The first is the largest, but the most impressive and scariest is the last one you pass before reaching Taft Point itself. There are well-developed exfoliation joints on either side of the clefts.

The joints that define The Fissures strike northeast at about 45 degrees. They are part of a set that defines the great northwestern face of Half Dome, the face of Nevada Fall, and the cleft between Mt. Broderick and Liberty Cap. Looking northeast from Taft Point, you'll see that the rock across the chasm is also riven by closely spaced northeast-striking joints. Why these joints are so deeply developed in this part of Yosemite is not known. Before leaving the point, be sure to examine the western cliff. It is controlled by a north-striking joint, and the grove of trees on the prow follows a parallel joint.

The northeast-striking joints in the cliff northeast of Taft Point are part of the set that defines the face of Half Dome.

Taft Point, looking northwest. A small length of the cliff is guarded by a railing, but the rest of the cliff is an unguarded 3,400-foot drop to the valley floor. The Rockslides is visible on the right.

If you paid attention to the rocks on this hike, you've noticed how complicated they are. The first part of the hike was on the 95-million-year-old Sentinel Granodiorite. As you emerged from the trees onto sandy ground around the point, you walked onto Taft Granite, which intruded El Capitan Granite. The El Capitan is darker than the Taft and contains conspicuous crystals of feldspar about 0.5 inch long. Both units are about 102 million years old. Much of the rock around Taft Point is El Capitan Granite, but the benchmark on the point is mounted in a thin dike of Taft Granite. Near the cliff edge just east of the point convoluted blobs of dark gray diorite are mingled with the Taft Granite.

Given the important role that joints play in shaping the landscape and controlling how water moves over and through rocks, it is natural to wonder what controls their orientation and spacing, and why they are arranged in parallel sets. One explanation that geologists have offered is that such joints form as crystallizing plutons cool and crack deep beneath the surface; however, in Yosemite the joints cut right across contacts between plutons of different ages, so that explanation doesn't work here.

The northwest-striking set that plays such a prominent role in the topography around Taft Point is nearly parallel to the major fault zone that bounds the eastern side of the Sierra Nevada. This is the westernmost fault zone of the Basin and Range province, the broad area of mountain ranges and

The Three Brothers as seen from Taft Point. This formation, which lies east of El Capitan and west of Yosemite Falls, demonstrates that not all steep joints are vertical. These dip to the west, toward Eagle Creek in the left of the photo.

valleys that stretches between the Sierra Nevada and the Wasatch Range and Colorado Plateau of Utah. The Basin and Range province overlies an area where Earth's crust is stretching and breaking, forming faults. The province has been expanding westward over the past several million years, and we might wonder if the northwest-striking joints around Taft Point are fledgling faults that are forming as the Basin and Range province bites its way westward into the Sierra Nevada block.

The prominent northeast-striking set that controls upper Yosemite Creek and the face of Half Dome is at about a 45-degree angle to those faults and requires a different explanation. They, too, may be young faults, in this case related to shearing of the Sierra Nevada block as the Pacific Plate slides to the northwest past the North American Plate. Because the Yosemite Creek joint, in particular, is so well developed and so long (at least 12 miles), it seems plausible that it is a fault. More study is needed.

11

Half a Dome Is Better Than None

SENTINEL DOME AND HALF DOME

See the map on page 130. Sentinel Dome lies about 0.5 mile southwest of the Glacier Point parking area. To get to the Glacier Point parking area from Yosemite Village, follow Northside Drive for 4.7 miles (from El Portal, follow California 140 east for 9.1 miles), then turn onto Southside Drive and follow it for 1 mile. Turn right onto California 41 south (Wawona Road), drive 9.2 miles, and turn left onto Glacier Point Road. The parking area lies about 15 miles down this road. You can reach the summit of Sentinel Dome by hiking about 1 mile from the Glacier Point parking area, with about 900 feet of elevation gain, or hiking about 1 mile from the Taft Point parking area (see vignette 10), with about 400 feet of elevation gain. The path to the top ascends the dome's relatively gentle northeastern ridge; though inviting, the dome's southern side is steeper than it looks and involves technical climbing. Sentinel Dome provides an excellent view of Half Dome, but few vantage points are as dramatic as the view of Half Dome from Mirror Lake (see vignette 7 for directions). Washburn Point, 0.5 mile south of Glacier Point, provides a unique view looking parallel to the northwest face of Half Dome.

Domes make up some of the most famous scenery of Yosemite National Park. Lembert Dome, Polly Dome, Fairview Dome, and others dominate the landscape of the high country around Tuolumne Meadows, and Half Dome, North Dome, Basket Dome, Mt. Starr King, and Sentinel Dome dominate the landscape of eastern Yosemite Valley. All provide fabulous views from their treeless summits. One might expect that the same process shaped all of these domes, but that isn't the case. Domes of the Tuolumne region owe their shapes to abrasion by glaciers (see vignette 13) and rapidly flowing slurries of sediment-laden water beneath the glaciers, whereas the domes of eastern Yosemite Valley owe their shapes largely to exfoliation (vignette 12). Sentinel Dome is smooth, rounded, and somewhat stained. It is clearly onionlike, composed of sheet upon sheet of granite, some thick, some thin. These layers represent classic exfoliation, with the sheets formed by jointing that parallels the topographic surface. In this vignette we examine Sentinel Dome and the evidence that tells geologists it is an exfoliation dome.

Although the walk up Sentinel Dome involves several hundred feet of elevation gain, it's easy and well worth the effort, as the view is one of the best you can get in Yosemite without a strenuous hike. Looking west down the Valley, craggy Cathedral Rocks stands opposite El Capitan, and if the day is clear you might see the Coast Ranges, 100 miles away. Yosemite Falls and the gentle upland valley of Yosemite Creek (vignette 3) lie due north of Sentinel Dome. Continuing to the right, Mt. Hoffmann (9 miles away) and Mt. Conness (22 miles away) dominate the skyline. North Dome (left) and Basket Dome (right) sit in front of Mt. Hoffmann. On the skyline to the right of Mt. Conness, beyond the bare rock of upper Tenaya Canyon, are Tenaya Peak, Echo Peaks, Cockscomb, and Clouds Rest, with Half Dome in front. Farther right, southeast of Sentinel Dome, is the beautiful Clark Range, with the barren domes of Mt. Starr King in front.

Let's examine the detective work that allowed us to state that Sentinel Dome is an exfoliation dome rather than one that was sculpted by ice. For comparison we offer Lembert Dome, which looms over eastern Tuolumne Meadows where the Tuolumne River crosses under Tioga Road. Both domes rise around 700 feet above the surrounding area, and both have a steep western face. Lembert Dome is about 1,300 feet higher in elevation than Sentinel Dome, but in gross form they are quite similar.

There are two obvious features that tell us Lembert Dome was scoured by ice in the recent geologic past: glacial erratics of diverse rock types (see vignette 14) dot the flatter parts of upper Lembert Dome, and much of the dome has been polished and striated. Sentinel Dome, in contrast, lacks glacial erratics, polish, and striations. Most of the boulders on top of Sentinel Dome are clearly the deeply weathered, crumbly remnants of exfoliation sheets that formed in place. A few of these boulders are rounded and look like erratics, but they are composed of the same rock as the bedrock of the dome itself. This is a bit of a tricky call because Sentinel Dome contains three different rock types—the El Capitan Granite, the granodiorite of Kuna Crest, and the Sentinel Granodiorite (named for exposures at Sentinel Rock to the west)—but because there are no boulders of rock types not present in the local bedrock, these boulders have probably weathered in place, rather than being brought here by glaciers.

There is no glacial polish anywhere on Sentinel Dome or in its vicinity. The lack of polish, combined with the brown and orange stains on the rock surface, give Sentinel Dome a look of antiquity that the Tuolumne domes lack. This staining tells us Sentinel Dome has been exposed and weathering much longer than the Tuolumne domes—hundreds of thousands of years versus 15,000 to 20,000 years. The stains are primarily the result of the oxidation of iron in the minerals in the granite, such as hornblende and biotite. In places this weathering process has produced an outer shell on the rock called a *weathering rind*. Typically about 1 inch thick, these rinds provide good handholds and footholds for climbing. Over tens of thousands of years, water flowing over the surface has eroded thin grooves and shallow pits in the rock. The staining, weathering rind, grooves, pits, and lack of erratics and polish all suggest that this dome has not been overtopped by ice for a long time, if ever. If it had, the ice would have scraped away the relatively weak features resulting from weathering, leaving behind polish and striations.

Sentinel Dome as viewed from the south (top), and Lembert Dome viewed from the southwest (bottom). Up close, compare the weathered, brownish, textured surface of Sentinel Dome to the smoother, gleaming, ice-sculpted surface of Lembert Dome. Ice flowed from right to left across Lembert Dome, a classic roche moutonnée.

Thick and thin layers of granite on the south side of Sentinel Dome, including a sheaf of particularly thin layers near the center of the photo. All of these onionlike layers are the result of exfoliation.

Weathering rind on the side of Sentinel Dome. Rinds like this can make good holds for climbing.

These grooves and shallow pits on the surface of Sentinel Dome were likely formed by water running down the side of the dome. Glacial striations, in contrast, are shallow, much more linear, and tend to be oriented horizontally rather than vertically.

One of the best indications that Sentinel Dome has not been glaciated—at least not for a long time—is the presence of weathering pans near the summit. Like weathering rinds, grooves, and pits, weathering pans form when granitic rock weathers over a long period of time, not by glacial erosion. They form where water collects in small natural depressions, which are slowly enlarged as the water chemically alters the rock, breaking it down to sandy material called *grus*. The pans are usually flat-bottomed and may have overhanging rims, and smaller pans often coalesce to become larger pans.

Although no one is sure of the exact process by which weathering pans form, one theory suggests that the most active environment for the chemical breakdown of the rock is the zone where water and air meet along the margins of the pans. Consider a depression that has a pool of water and one or more small islands of rock in it. The most active weathering occurs along the outer shoreline of the pool and the margins of the islands. As the rock weathers, the overall shoreline grows outward, and the islands shrink as their shorelines grow inward. Eventually the islands no longer poke above the surface of the pool and thus cease to weather rapidly. If the water level drops, though, the islands reappear and begin weathering again. A kind of equilibrium is reached when the entire bottom of the pan is equally wet or dry at the same time. Thereafter, the entire pan weathers downward at the same rate, maintaining a relatively flat bottom. The sandy weathered material produced by this process is removed by wind, although some of it may remain in deeper pans.

Sentinel Dome offers a great view of Half Dome, probably the most famous granite monolith in the world. Sentinel Dome and Half Dome share many characteristics: both are rounded in form due to numerous exfoliation joints, both poked up above the upper reaches of the glaciers of Pleistocene time, neither has glacial erratics on its summit, and both are deeply weathered. However, Half Dome is different in one obvious way, and no book on Yosemite geology would be complete without an explanation for why Half Dome looks the way it does. In fact, the most common geological question asked of Yosemite interpretive rangers is "What happened to the other half of Half Dome?" It's a good question, but it's somewhat flawed because there really isn't a missing half. Half Dome is a bit of an optical illusion. When viewed from the floor of Yosemite Valley, it does appear as though roughly half of the dome is missing. However, when viewed from other points it looks much fuller. Washburn Point off Glacier Point Road is one of the best and most accessible viewpoints for seeing this, but even from Sentinel Dome it is clear that, rather than being truly half of a dome, Half Dome is actually a narrow ridge. The ridgeline is oriented northeast-southwest, with sheer walls on both its southeast side and its northwest side—the famous northwest face.

Large, coalescing weathering pans near the summit of Sentinel Dome.

Half Dome as seen from the summit of Sentinel Dome.

When viewed from this point on its southwest shoulder, Half Dome actually appears more like a full dome.

The northwest portion of the dome has obviously been cleaved off, but most of the dome remains. The late Yosemite geologist N. King Huber estimated that nearly 80 percent of the original dome is still intact.

So how did the sheer northwest face of Half Dome form? It may have formed during a colossal rock avalanche, as early geologists proposed, but more likely it developed incrementally. The northwest face occurs along a prominent northeast-southwest-trending, vertically oriented joint. This is a common orientation for joints in Yosemite (see vignette 12), and Tenaya Canyon developed along this same joint system. As glaciers repeatedly flowed down Tenaya Canyon, they probably quarried back the lower portion of the northwest face, leaving unstable rock overhangs. Exfoliation joints riddle the northwest face, and these would have further destabilized the overhangs. As the overhangs collapsed, the face grew sheer, and glaciers easily transported the rockfall debris down-valley. Although glaciers no longer occupy Tenaya Canyon, exfoliation joints continue to spawn rockfalls, shaping the northwest face of Half Dome (see vignette 5) and keeping it steep and beloved of rock climbers.

The color, cragginess, and baldness (well, lack of glacial erratics anyway) of Yosemite's unglaciated domes are reminiscent of the qualities that distinguish the old from the young among humans. Keep this in mind and, with a little practice, you'll find it easy to tell the relative ages of Yosemite's domes.

12

The Earth as an Onion

EXFOLIATION JOINTS

Visitors to Yosemite National Park cannot help but notice the way granite breaks into thin sheets parallel to the ground surface. This may seem a bit incongruous, given how much we've talked about the strength of Yosemite's granite, yet this process gives shape to Yosemite's domes, such as Sentinel Dome and Half Dome (vignette 11), North Dome, and others. The domal shapes develop as exfoliation joints form and the thin, curving sheets of rock between them break up and slough off the sides of the domes (exfoliation joints are also called *sheeting joints*). The term *exfoliation*, used in medicine, biology, and many other contexts, refers to the shedding of thin layers from the surface of an object. That clearly describes what these domes are doing.

This process has fascinated scientists for over a century, and many explanations have been put forth to explain why these joints form, including erosion, daily heating and cooling, shrinkage as a pluton cools shortly after it crystallizes, chemical weathering, and mechanical agents such as fire and frost. We can go ahead and put one of these explanations to rest. Exfoliation joints occur in many different types of rock, not just igneous rocks, ruling out the possibility that they only form as magma cools and becomes a pluton.

Before we discuss other explanations for how the joints form, let's take a look at the oft-photographed roadcut west of Yosemite Creek, which is a particularly good example of exfoliation jointing. Here the granite sheets have a relatively small range of thicknesses (about 1 or 2 feet) and are parallel to the surface of the outcrop, which makes them slightly convex. You may see water leaking out of the joints, especially during early summer. This is groundwater that flowed along the joints until it found its way out. Exfoliation joints thus play an important role in the movement of groundwater through otherwise impermeable granite. These joints also make it easier for glaciers to erode bedrock by plucking, and they play a key role in controlling the location and magnitude of rockfalls. For these reasons, exfoliation joints merit close study.

When a sheet of rock is exposed at the surface, cracks form that run perpendicular to the exfoliation joints, and weathering agents exploit them to break the sheets apart. For example, plant roots grow in the

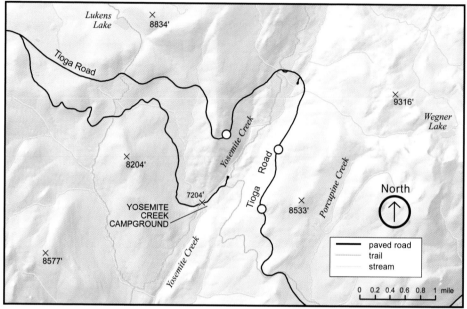

Places along Tioga Road (marked by white circles) where exfoliation joints are especially well displayed in roadcuts.

GETTING THERE

Exfoliation joints occur all around the world, but those in Yosemite National Park are perhaps the best known. Though they are prominent in roadcuts along much of the eastern part of Tioga Road (California 120), a particularly impressive exposure occurs 1.3 miles west of the Yosemite Creek crossing. To get there from Big Oak Flat Entrance Station, follow California 120 east (Big Oak Flat Road) for 7.8 miles. Turn left, continuing on California 120 east (Tioga Road) for approximately 18.4 miles (3.9 miles east of the turnoff to White Wolf Campground). To get there from Tioga Pass Entrance Station, follow California 120 west (Tioga Road) approximately 28.1 miles. Park in the small parking area on the south side of the road and walk a short distance to the top of the outcrop. Other excellent places from which to view exfoliation joints include a large paved pullout about 1.1 miles southeast of the Yosemite Creek crossing and a similar pullout about 0.75 mile farther south. Olmsted Point is also a superb viewing point, especially if you leave the parking lot and make the short hike to Olmsted Point proper (vignette 14).

An outstanding example of exfoliation joints exposed along Tioga Road west of the Yosemite Creek crossing. These slabs are particularly uniform in thickness but thicken slightly the farther they are from the outcrop's upper surface. This thickening with depth is a common feature of exfoliated granite, but why it happens is not well understood. Joint-bounded slabs that have broken out of the outcrop have such sharp, clean lines that they look artificial.

cracks and pry the rock apart, as does water when it expands as it freezes in the cracks. With enough time, water chemically breaks the rock down and carries it away, enlarging a crack until it cuts through a sheet. In most cases, multiple factors working together break the sheets apart. It's common to see plates of rock on the surface of Yosemite's domes that clearly would fit back together if reorganized a bit. These smaller pieces can slide downhill, exposing a new sheet beneath. On some domes there are smaller plates on the top but none on the steeper flanks. Instead, those plates rest in piles at the bottom of the slopes.

Gently dipping joints such as these have long been called *unloading joints* because geologists thought they formed as erosion "unloaded" rock from the surface. Since granite forms a few miles or so beneath the surface, it is under enormous pressure equivalent to about two thousand or three thousand of our atmospheres, or that experienced 10 to 20 miles deep in the ocean (if the oceans actually got that deep, which they don't). In this hypothesis, as erosion slowly removes the rocks above the granite,

The 10-inch-thick plate upon which the second author is standing, near The Rostrum, measures about 3 feet square. It clearly has moved at least twice from where it was attached to the plate on the left, as shown by its ghostly outline on the surface beneath. How did it move on such a gently dipping surface? Perhaps earthquake shaking caused it to move. If the plate has been detached for 100,000 years (not a bad guess, but no more than a guess), then even if large earthquakes such as the 1872 Owens Valley shock (see vignette 7) only shake the area every 1,000 years, that's one hundred chances for the plate to be shaken and move at least a bit.

Intense exfoliation jointing defines the sloping southern side of Mt. Hoffmann.

Deeply eroded remnants of sheets on the flank of Sentinel Dome. At some point in the past, the surface of the dome was at least as high as the top of this stack. Rocks on each side of the stack either slid off the side of the dome or eroded completely into sandy debris that was transported away by water or wind.

the vertical pressure decreases, but the rock is still held rigidly on the sides. The rock responds by popping up vertically, opening along joints parallel to the surface.

This explanation seems reasonable, and it has been put forth as the answer in textbooks for decades. However, it has been questioned for at least a few reasons. First, if unloading alone were the cause of the joints, then deeply eroded rocks everywhere should have them, but they don't. Second, the relief of vertical pressure, by itself, doesn't cause vertical tension. Tensile stress, which is tension related to the stretching of something until it is taut, is necessary for a joint to open vertically.

Recently, professor Stephen Martel of the University of Hawaii proposed another hypothesis explaining the origin of exfoliation joints. He noted that the joints are prominent when the following conditions are satisfied: the rock surface is convex, the rock is strong (as granite is), and the rock is under strong horizontal compression—that is, it is being squeezed hard from the sides. The sideways compression pushes a convex rock surface upward. The rock deep below the surface reacts to the compression by pulling down, creating tension perpendicular to the convex surface of the rock near Earth's surface. If the surface is sufficiently convex and the sideways compression strong enough, the resulting perpendicular tension overcomes both the force of gravity and the tensile strength of the rock, and joints pop open parallel to the convex surface.

The mathematical description of this process is involved, but the basics can be demonstrated with a simple experimental apparatus, which can then be eaten. Take a sturdy slice of bread with the usual convex top (a good sourdough works well). With a sharp knife, slice a 1- or 2-inch-long slit parallel to, and about 0.5 inch from, the top of the slice. Lay the slice on a flat surface and gently apply pressure from the sides with your hands, compressing the slice. The slit will open. This makes it easy to see how compression facilitates the opening of exfoliation joints: some of the horizontal motion of your hands is transferred into motion parallel to the curving top of the bread, and this exerts an upward tug. If you find

An edible example of the formation of exfoliation joints. The upper slice has two curved slits cut into it, parallel to the convex upper surface. The slits open when the bread is squeezed from the sides. This demonstrates that the combination of horizontal compression and a convex upper surface generates vertical tensile stress.

a piece of bread with a concave edge, cut a slit parallel to it and squeeze the sides of the slice. The slit will stay closed, illustrating the importance of a convex surface in the formation of exfoliation joints.

Our example is cooked up (as it were) because we put a slit in the bread to begin with, but this is required because gluten-rich bread has different physical properties than rock. Instead of cracking, it stretches and tears. Nevertheless, our model shows how squeezing convex rock bodies can cause joints to form that parallel the rock surface—joints that eventually lead to the interesting landscape we see today in Yosemite. Regardless of how these joints form, exfoliation is ongoing as convex rock surfaces shed onionlike layers in Yosemite.

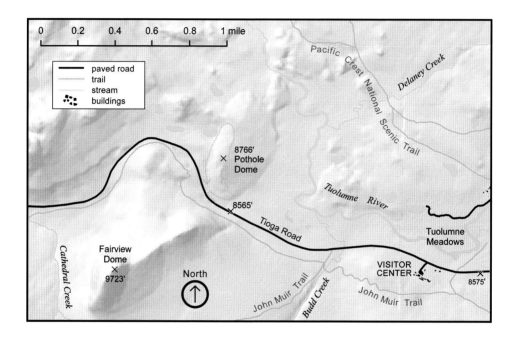

GETTING THERE

Pothole Dome is located just north of California 120 (Tioga Road), at the western end of Tuolumne Meadows (1.1 miles west of the entrance to the Tuolumne Meadows Visitor Center). To get there from Big Oak Flat Entrance Station, follow California 120 east (Big Oak Flat Road) for 7.8 miles. Turn left, continuing on California 120 east (Tioga Road) for approximately 37.4 miles. To get there from Tioga Pass Entrance Station, follow California 120 west (Tioga Road) approximately 9.2 miles. Park in the signed parking area on the north side of the road. We describe a short, mostly gentle loop over the top of Pothole Dome. Be sure to follow the trail, and do not cut across the meadow.

13

The Ice Went Thataway!

THE SHAPING OF POTHOLE DOME

Tuolumne Meadows is known for its smooth granite domes and delicate spires. This striking landscape was sculpted by ice, and the proof is clear to anyone willing to walk a bit and ponder the evidence. Less well-known is that many of the minor details on some of the domes were carved by rapidly flowing rivers that flowed under the ice, not by the ice itself. Although it is neither as tall nor as visited as its nearby and more famous cousin Lembert Dome, Pothole Dome displays features of both sculpting forces, and alert detectives can determine the direction the ice flowed and the maximum height it achieved. The view from the top of the dome is outstanding, too.

Pothole Dome is part of a northeast-trending ridge that includes tall Fairview Dome (9,723 feet) to the south and smaller unnamed domes to the north near the Tuolumne River. Tioga Road and the parking area for Pothole Dome are located in a gap in this ridge. The top of the dome is not quite visible from the parking area, and it is difficult to appreciate that the dome rises 200 feet above the meadow, to an elevation of 8,766 feet.

Viewed from the parking area, the dome presents a long, gentle right-hand (eastern) slope and a much steeper left-hand (western) slope. This asymmetry is important to interpreting the direction the ice flowed. From the parking area follow the trail, which runs along the side of the road to the left until it meets the southern edge of the dome and swings sharply right. Pothole Dome is carved in the Cathedral Peak Granite, broken boulders of which are piled up next to the trail just below the road. The fresh broken surfaces of these rocks contain obvious, large pink crystals of orthoclase and small crystals of white plagioclase, clear gray quartz, and black biotite. Rocks on the dome are mostly polished and look quite different, as we will see. Follow the trail around the western edge of the meadow to the base of the dome, and then to the right (east) to the trees at the southeastern edge of the dome, where you can leave the trail and begin your ascent.

The first clue that ice once covered Pothole Dome is that much of its surface is highly polished. There are only a few natural processes that can polish a rough rock surface, and scouring by grit-laden ice is the most common (see the introduction for information about glacial processes).

In places the polish has been eroded away, revealing the rough granite underneath. Although most of the dome is composed of Cathedral Peak Granite, thin, light-colored, sugary-looking aplite dikes, composed of quartz and feldspar, crisscross the dome, and they commonly preserve polish better than the coarser-grained granite.

Glacial striations are another sure sign that a glacier scoured a rock surface. These are grooves that rocks embedded in the underside of the ice scratched into the underlying surface. The overlying ice can be hundreds of feet thick, and the weight of this mass turns these rocks into powerful agents of erosion.

The southeastern end of Pothole Dome is striated, but many of the striations are subtle. Other forces, such as rivulets of running water, can produce grooves in rock, so it is important to be able to distinguish glacial striations from grooves formed by other processes. Grooves here parallel the long axis of the dome surface, toward its top. Grooves carved by water tend to run down the steepest surface because water generally (but not always) flows downhill by the most direct line possible. These striations run at a significant angle to the steepest downhill direction, indicating they have a glacial origin.

Glacial polish has been preserved on this aplite dike, but the granite around it has been weathered and is no longer smooth. Striations are visible in the dike, cutting across it from left to right. Knife for scale.

Chatter marks, a few feet across, on the southeastern side of Pothole Dome. Glacial striations run perpendicular to the arcs of the chatter marks. Pen (center) for scale.

Now that we know these grooves were carved by rock embedded in glacial ice, can we tell which direction the ice flowed? The chatter marks low on the southeastern side of the dome give us a clue. If you know how to interpret their shape, these crescent-shaped depressions will tell you which way the ice flowed. Note that the chatter marks are all facing the same way: convex uphill. We'll discuss this interesting phenomenon a bit later in this vignette.

It seems clear that the ice could only have flowed parallel to the striations because they were carved by rocks and grit carried by the ice. That gives us two possibilities: either the ice flowed up the slope toward the summit of the dome, in the convex direction of the chatter marks, or down the slope, toward Tuolumne Meadows, in the concave direction. Surely it must have flowed downhill, because the slope is steep—10 degrees. It would take a lot of work to push ice up that slope. Let's walk to the top of the dome to test this hypothesis.

Along the way you'll see abundant striations and polished surfaces and that the clean granite surface of the dome is littered with boulders up to 6 feet or more in diameter. How did they get here? Several possibilities come to mind: they could have fallen from higher ground or been carried here by rivers or ice, or they may merely be eroded pieces of the rock the dome is composed of; such remnant rocks left behind as surrounding rock weathers and washes away are called *lag boulders* or *core-stones* (vignette 21).

We can rule out the first possibility because the boulders persist all the way to the top of this and most of the other domes in the Tuolumne Meadows region, so there is nowhere they could have fallen from. The second possibility seems highly unlikely, because it would take quite a river to carry 6-foot-diameter boulders to the top of the dome. That leaves ice and in-place weathering as possible origins.

How do you determine which of these origins is most likely? One way is to see what the boulders are made of. If they differ from the underlying bedrock, they must have been carried here. If they match the bedrock, then an in-place weathering origin is likely. Another way is to see if the boulders sit on polished surfaces. If so, they must be erratics, carried to their present locations by glacial ice. Although most of the boulders on Pothole Dome look like Cathedral Peak Granite, which makes up the dome, some of them differ in detail. These boulders are composed of Half Dome Granodiorite and the granodiorite of Kuna Crest, which have smaller orthoclase crystals and a higher proportion of the dark minerals hornblende and biotite than Cathedral Peak Granite. These boulders are indeed glacial erratics.

An erratic composed of Half Dome Granodiorite, showing a dark enclave, conspicuous hornblende crystals about 0.5 inch across, and no large crystals of orthoclase, which are clearly visible in the underlying Cathedral Peak Granite that makes up Pothole Dome. These distinguishing characteristics demonstrate that this boulder is from somewhere else. Pen for scale.

Most of the boulders on the lower slopes of Pothole Dome are Cathedral Peak Granite, but as you approach the shallow bowl that lies just south of the true summit, look for a group of erratics, each about 4 feet across, near a prominent aplite dike. Many of these boulders are Half Dome Granodiorite. Although from a distance this granodiorite superficially resembles the Cathedral Peak Granite, closer inspection reveals that it has prominent black, rectangular crystals of hornblende up to 0.5 inch across. Some of the erratics also contain a blob or two of dark rock. These dark enclaves (see vignette 1) are quite rare in the Cathedral Peak Granite but common in the Half Dome Granodiorite. The nearest outcrops of the Half Dome Granodiorite are about 1.5 miles to the west and 5 miles to the east.

The top of the dome, which is littered with hundreds of erratics, is a good place to view the grand landscape of Tuolumne Meadows. The nearby, rounded dome to the south, just across the road, is the northeastern shoulder of Fairview Dome; Fairview itself is the taller, pointier dome to the right. Just to the left of Fairview Dome, Cathedral Peak peeps over the tree-covered ridge. The sharply pointed, distant peak to its left is Cockscomb, and the pointy, clustered group to its left is Unicorn Peak. Because they are pointed, these peaks must not have been overtopped by ice (which would have worn them down), but the smoothness of the landscape beneath them is evidence that everything else was covered with ice. If you can visualize an imaginary line extending out across the meadows from the boundary separating the formerly glaciated landscape below from the unglaciated rock on the peaks, it is clear that the ice was some 1,800 feet thick over the meadows.

Having established that the boulders were deposited by ice, our remaining task is to verify the direction the ice flowed. The erratics could provide a clue if we could determine their source, but the Half Dome Granodiorite occurs west of here (such as near Tenaya Lake) and east of here (near Tioga Pass). The granodiorite of Kuna Crest also occurs east and west of here, so the erratics are of no help. Upon reaching the summit of the dome, however, you should realize that our original hypothesis—that the ice flowed east down the slope you just walked up—must be wrong. In order to have flowed down the eastern side, the ice would have had to surmount the much steeper western side, which is nearly a cliff.

In fact, glaciers respond to topographic slope in an entirely different manner than rivers. Rivers always flow downslope, and if they encounter a rise, the water ponds until it spills over or around the rise and flows downslope again. Glaciers, in contrast, are able to flow up modest topographic slopes. Like rivers, glaciers flow downslope due to gravity, but the direction of flow is determined by the surface slope of the glacier itself, not necessarily the slope of the topography over which it flows. So which slope of Pothole Dome did the ice flow up?

The strongly asymmetric profile of the dome provides us with a key piece of evidence. This dome, with one steep side and one not-so-steep side, is a classic glacial landform called a *roche moutonnée*. Ice flowed up the gentler side and over the top, plucking blocks off of the lee (downstream) side to produce the steep face. As noted by Bob Sharp in *Living Ice*, "The smoothed, gentler flank provides the name, a *moutonnée* being the smoothly curled wig worn by barristers and judges in early European and British courts. Some people prefer an analogy to the back end of a sheep (*mouton*)." Thus, you are looking at a "rock wig" or "rock sheep."

An aerial view of this region shows that most of the domes in the Tuolumne Meadows area are asymmetrical, with steep western sides and much gentler eastern sides. The shape of Pothole Dome and its other glacial markings, as well as other evidence in the park, tells us that the thick ice sheet in Tuolumne Meadows generally sloped down to the west and flowed east to west across this landscape as it sought the lower ground of the Tuolumne River and Tenaya Canyon. The large volume of ice easily rode up and over "small" topographical features such as Pothole Dome. By such reasoning here and elsewhere, we can deduce that the convex side of crescent-shaped chatter marks point in the direction that the ice flowed. They likely formed as large rocks embedded in the ice impacted the bedrock surface, exerting pressures that cracked the bedrock into characteristic crescent shapes.

After peering over the western edge from the summit, walk north (to the right) and then down the western side toward the small pond. There are several routes, and by walking first north and then west you can find one gentle enough to suit your needs. Don't go this way if you feel uneasy or if the rock is wet or icy; just return the way you came.

Lodgepole pines and a moraine-bounded pond occupy the glade at the bottom of the steep western side of Pothole Dome. Walk to the left (south after descending the dome) along the faint trail near the base of the cliff. After a few hundred yards the trail climbs a small rise of loose rocks and dirt. Be sure to stay close to the dome and not veer to the right, onto a more prominent trail. The boulders and cobbles in this rocky deposit are of many rock types and are mostly quite rounded, like the rocks found in a river. But the Tuolumne River flows to the north in this area, around Pothole Dome and the smaller surrounding domes. And Tioga Road climbs 100 feet west from the Pothole Dome parking area before descending toward Tenaya Lake. Thus, it seems unlikely that a major river ever flowed through the gap south of Pothole Dome. So where did these river cobbles come from?

Continue south, toward the parking area, staying close to the base of the dome; eventually you rejoin the path you started on. After walking a short distance east along the exposed southern edge of the dome, walk up onto the dome's flank until you find a set of deep potholes carved in the granite.

View of Pothole Dome from the parking area in the fall, when the meadow is brown. Like many domes in the Tuolumne Meadows area, Pothole Dome has an asymmetric profile, with a long, gentle slope on the east (right) *and a much sharper slope on the west* (left). *This classic glacial landform is called* roche moutonnée.

This side of the dome is a bit steep, so be careful and avoid the polished areas. Stay away if the rock is wet.

Potholes are deep cylindrical holes in solid rock. They form when fast-flowing water swirls rocks around in a tight circle in the same place over a long period of time. The stones become trapped in the initial depression they drill into the rock, which allows them to continue to erode the spot and deepen the holes. As the original stones are worn away, new stones fall into the pothole, continuing the process.

Several potholes occur on the southern flank of Pothole Dome. In the larger ones you can find highly rounded boulders up to several feet in diameter. These boulders were polished by the swirling action of sediment in the pothole or as they were swirled around the edges of the pothole. These boulders were clearly the tools that did the drilling. But what river was powerful enough to move them? And how could a river have flowed over the surface of the dome? These potholes caught the attention of pioneering geologist G. K. Gilbert over a century ago. Building on suggestions by others, he proposed that the potholes formed as water flowed down through the glacial ice and encountered the bedrock of the dome. He thought the process was similar to the way deep plunge pools can be excavated at the bases of waterfalls by the force of falling water.

Shallow channels carved into the polished rock surface of Pothole Dome suggest that these potholes were actually carved by water flowing along the rock-ice interface beneath the glacier rather than by water that sank through the ice surface. How could water do this? The most likely explanation is that the weight of the overlying ice pressurized the water, allowing it to flow uphill and swirl through the potholes. The larger boulders in the potholes are 3 to 4 feet in diameter, and it would have taken a strong current to move them.

Pothole Dome's potholes are an excellent place to ponder the power of subglacial rivers, and nice places for trees to grow on otherwise inhospitable bare rock. This is the most conspicuous pothole on the south side of Pothole Dome. Note the large, polished boulder of light-colored Johnson Granite Porphyry, about 4 feet across, in the foreground and partially obscured by shadow.

This large pothole is located in upper Tenaya Canyon.

Thus, at Pothole Dome we see evidence of the work of both rivers and glaciers. Interestingly, this work was probably done more or less at the same time. Glacial ice planed over the dome, polishing the less-steep side and plucking blocks from the other side, making it steeper. Meanwhile, meltwater flowed along the base of the glacier, driven in part by the intense pressure of the overlying ice. It collected in rather large rivers that sculpted the underlying bedrock, scoured the potholes, and deposited the rounded cobbles at the western base of Pothole Dome.

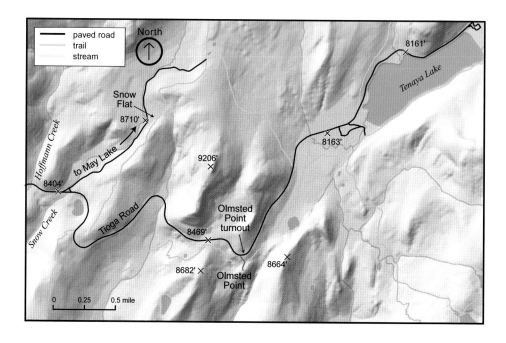

GETTING THERE

Olmsted Point is located along California 120 (Tioga Road), 1.5 miles west of Tenaya Lake. To get there from Big Oak Flat Entrance Station, follow California 120 east (Big Oak Flat Road) for 7.8 miles. Turn left, continuing on California 120 east (Tioga Road) for approximately 29.3 miles. To get there from Tioga Pass Entrance Station, follow California 120 west (Tioga Road) approximately 17.2 miles (9.2 miles west of the Tuolumne Meadows Visitor Center). Turn left into the large paved parking area on the south side of the road. Though the parking area is often referred to as Olmsted Point, the actual Olmsted Point, a bedrock dome south of the parking area, is reached by a short, 0.25-mile walk on a good trail that starts on the south side of the parking area.

14

Exotic Erratics
GLACIALLY TRANSPORTED BOULDERS AT OLMSTED POINT

The large paved parking area off Tioga Road at Olmsted Point is a popular viewpoint, and for good reason. It offers great views of the high country, including a unique perspective on Half Dome. The bare bedrock slabs near the parking area are a great place for kids to burn off energy after a long drive. And Olmsted Point showcases some impressive geology, especially in terms of the effects of the Tioga glaciation. There are neat things to see in the immediate vicinity of the parking area, but it can be crowded. A short walk on the trail south to Olmsted Point proper is enjoyable for the geology, views, and relative solitude.

During glacial periods, the vast ice field covering the Tuolumne Meadows area was so thick that on its southwestern margin ice spilled over from the Tuolumne River drainage basin into the headwaters of Tenaya Creek, which is part of the Merced River drainage basin. The ice swept over the divide separating these major drainage basins, smoothing Pywiack and Polly domes and carving out the basin now occupied by Tenaya Lake. From there, the glacier flowed across Olmsted Point and down Tenaya Canyon into Yosemite Valley, where it scoured the northwest face of Half Dome (see vignette 11) and joined the glacier in the Valley just west of Half Dome. Abundant glacial polish and striations provide evidence of this glacial path.

As the climate warmed during the end of the Tioga glaciation some 15,000 years ago, the glacier that occupied Tenaya Canyon retreated, leaving behind till and erratics. Till is a jumbled mix of silt, sand, gravel, and boulders. Erratics are boulders that were transported by ice, often over long distances, and left behind as the ice melted. The boulders may have been plucked from bedrock beneath the glacier, or they may have fallen from cirque headwalls and ridges projecting above the ice surface. Once embedded in the ice, the boulders were transported, conveyer-belt-style, down Yosemite's valleys and deposited as the ice melted. Erratic boulders may be solitary, appearing as single boulders on a sea of smooth bedrock, or they may be clustered in groups of dozens or even hundreds.

The erratics at Olmsted Point are mostly solitary boulders. It may be that the retreating glacier left the boulders exactly as they appear today,

Glacial erratics on the jointed, polished, and striated bedrock surface adjacent to the Olmsted Point parking area. —Courtesy of Garry Hayes

or perhaps they were once surrounded by heaps of till made up of finer-grained sediment that has since washed away. (Your authors have experienced intense summer thunderstorms in this area and witnessed a lot of sediment being transported by the resulting sheets of water pouring off the domes.)

The boulders in the immediate vicinity of the Olmsted Point parking area are erratics in the sense that they clearly were brought here by glaciers. How do we know this? In some cases the boulders rest on bedrock surfaces that are polished and striated. Such surfaces have a glacial origin, and boulders resting upon them almost certainly have a glacial origin as well, rather than being, for example, core-stones that weathered in place (see vignette 21). Erratics are common on the glacially polished and striated domes in and around Tuolumne Meadows. Rock type is another clue that can be used to determine if a boulder is an erratic. In many cases erratics are composed of a different rock type than that of the bedrock surface they rest on. This is a good indication that they were transported by a glacier and set down in a new location. However, most of the erratics at Olmsted Point are not what we might call "classic" erratics because they are composed of the same rock type as the rock they rest on: the Half Dome Granodiorite. That tells us that these boulders didn't travel very far, as the eastern boundary of the Half Dome Granodiorite pluton is just east of Tenaya Lake. If the glaciers were moving westward and plucked the boulders at the eastern edge of the pluton, they would have traveled at most 2 to 3 miles before being dropped at Olmsted Point. If you take the short trail to Olmsted Point proper, however, you can find scattered boulders of the granodiorite of Kuna Crest, which is considerably darker than the underlying Half Dome Granodiorite. These rocks originated many miles to the east. Careful sleuthing of the bedrock slabs across the road from the parking area may even reveal an erratic or two of tan to brown metamorphic rock transported from near May Lake, about 3 miles to the north (vignette 17).

You can also find erratics of the more "classic" variety along the western shore of Tenaya Lake (a great place to go for a swim—the surface water is warmer than you'd think) where there is a picnic area. Here you will find boulders of Cathedral Peak Granite sitting on Half Dome Granodiorite. The Cathedral Peak Granite boulders are easy to distinguish from the underlying Half Dome Granodiorite because they have large pink orthoclase crystals, which from a distance make the boulders look knobby or warty. These boulders were derived from the bedrock between the east shore of the lake and Tuolumne Meadows.

Erratics at Olmsted Point proper. The boulder in the foreground is composed of the granodiorite of Kuna Crest, which is clearly much darker than the underlying Half Dome Granodiorite.

Although this boulder of Cathedral Peak Granite is composed of the same rock as Pothole Dome, on which it rests, it sits on a glacially polished and striated surface, indicating that it is a glacial erratic. Ice flowed toward your first author.

Most of the erratics you are likely to see in the Tuolumne Meadows area and along Tioga Road were deposited during the waning stages of the Tioga glaciation; therefore, they appear relatively fresh, having competent, unweathered surfaces. Older erratics may be deeply weathered, to the point where individual crystals can be swept from the rock surface by hand. Erratics from earlier glaciations are harder to find, mainly because weathering and erosion have destroyed them. On the rare occasions that these older erratics are preserved, they sometimes have a remarkable appearance.

A surprising number of older erratics on bedrock domes are perched, sometimes precariously, on pedestals of bedrock. Obviously these erratics were not set down on the pedestals, because a glacier would have easily ground off such a protrusion; nor did the pedestals grow upward under the boulders. The pedestals formed gradually underneath the boulders as the surrounding bedrock surface was lowered. How did this happen?

The key to pedestal formation is exfoliation. In vignettes 11 and 12 we describe how bedrock surfaces, especially domes, erode by exfoliation, the sliding away of thin shells of rock along joints that parallel the surface.

Exfoliation joints form due to internal stresses in the rock, and other forces, primarily involving water, cause the slabs of rock detached along exfoliation joints to disintegrate in place or move downslope. Water accelerates the weathering of these bedrock slabs, and when it freezes in the joints it exerts an outward force that can ratchet the slabs downslope.

Why, you may ask, doesn't the pedestal itself exfoliate? There are two possible explanations. First, the boulder may act as an umbrella of sorts, protecting the rock underneath it from the weathering effects of rain, ice, and sun (solar heating). The second and more likely explanation is that the sheer weight of the boulder clamps the exfoliation joints in a pedestal closed, prohibiting water from entering them. This eliminates the most effective mechanism for weathering and eroding the jointed rock. Even if water did penetrate the joints and freeze, its expansion probably would not exceed the downward force of the weight of the boulder. Thus, a pedestal grows because the weight of the boulder prevents the agents of weathering and erosion from destroying it. With time, the boulder becomes perched higher and higher as the rock surface around it is lowered.

A perched erratic on Moraine Dome near Little Yosemite Valley. This boulder of knobby Cathedral Peak Granite rests on a pedestal of Half Dome Granodiorite that is about 3 feet tall. The rock pedestal suggests that the erratic was set down during an earlier glaciation—a glaciation that happened long enough ago that the rock surface around it has since been lowered 3 feet.

In addition to marking the extent of earlier glaciations in spectacular fashion, perched erratics are also useful for assessing the limits on the amount of ground shaking that occurred during prehistoric earthquakes. Even though earthquakes have triggered huge rock avalanches in Yosemite Valley (vignette 7), they were unable to topple these precariously perched boulders. That's a good thing, because as dating techniques become more sophisticated, it may someday be possible to determine when these perched erratics were laid down by glaciers, providing insight into the early glacial history of the park. In any case, these ancient perched boulders are certainly some of the most fascinating geologic features in Yosemite.

Not all perched boulders are erratics; core-stones or other locally derived boulders can also become perched through the same process. Perhaps the most spectacular of its kind, this perched boulder was discovered in the Tuolumne River drainage by National Park Service staff in 2006. The boulder weighs some 30 tons and sits precariously on a narrow pedestal that is 3 feet tall. Its origin has not been determined. Another perched boulder can be seen in the distance. —Jim Roche photo, courtesy of the National Park Service

15

Why Are There Trees Poking Out of Tenaya Lake?

THE GREAT MEDIEVAL MEGADROUGHT

One of the best ways to understand what Earth's climate might be in the future is to study what it was in the past. The study of past climates is known as *paleoclimatology*, and scientists working in the field, called *paleoclimatologists*, look at many pieces of evidence when determining the character of an ancient climate: data recorded by instruments (which go back 100 to 200 years at best), historical records (which go back several thousand years), tree rings (which extend the record to over 10,000 years), ice cores (extending back 100,000 years or more), ocean sediment cores (extending to tens of millions of years), and fossils (which can take the record back hundreds of millions of years). Although the information is less abundant and less precise as one goes back in time, paleoclimatological studies have given us a good picture of Earth's overall climate for the past several hundred million years, revealing that Earth's climate has changed substantially and often over geologic time.

These studies are of special interest in the American West owing to the scarcity of water. Anything that changes the amount of precipitation, or whether it falls as rain or snow, will have a profound effect on the people, animals, and plants of the West. One of the more frightening findings of paleoclimatological studies is that western North America has suffered through remarkably deep and prolonged droughts, or megadroughts, at least a few times in the past 1,000 years. If a megadrought developed again, the West would suffer a water deficit so great that much of the population would have to either move elsewhere or import or desalinate water using massive amounts of energy and money. Evidence for these ancient droughts is found in a variety of places, but some of the strongest evidence is found in the heart of Yosemite National Park, poking up out of Tenaya Lake.

Before getting to that evidence, let's talk about the lake itself. Much of the information below comes from Ned Andrews, a U.S. Geological Survey hydrologist who has been studying the lake for several years. Tenaya Lake, with an area of 222 acres, has beaches on its northeastern, western, and southwestern shores that make it a popular for picnicking. The

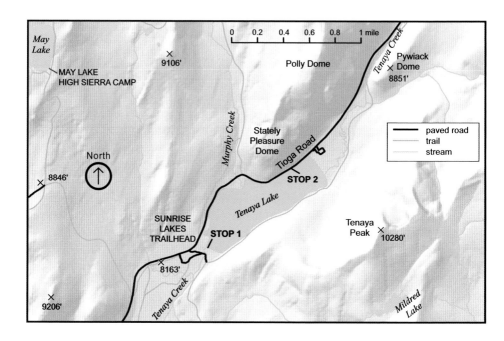

GETTING THERE

Tenaya Lake is located along California 120 (Tioga Road), near the center of the park. To get there from Big Oak Flat Entrance Station, follow California 120 east (Big Oak Flat Road) for 7.8 miles. Turn left, continuing on California 120 east (Tioga Road) for approximately 30.8 miles to the lake's outlet. To get there from Tioga Pass Entrance Station, follow California 120 west (Tioga Road) approximately 15.8 miles. Although parking is available at several places along the lake, in this vignette we describe two stops. Stop 1, the lake's outlet, is reached by a walk of a few hundred yards down the Sunrise Lakes Trail from either of two parking areas near the outlet. The outlet stream crosses the Sunrise Lakes Trail about 300 feet downstream from the lake's edge. Stop 2 is the lake's midsection, visible from parking spots along Tioga Road.

Looking west across Tenaya Lake from the north flank of Tenaya Peak. The bare slopes of Stately Pleasure Dome come right down to the lake in the center of the photo, with Mt. Hoffmann (left) *and Tuolumne Peak* (right) *on the skyline.*

water is chilly, but in late summer the surface water is warm enough for swimming. Tenaya Lake is fed by three principal sources: Tenaya Creek, Murphy Creek, and a branch of Tenaya Creek that drains Cathedral Lakes. The total catchment area, the area feeding these streams, is about 5,200 acres.

At stop 1, Tenaya Lake drains to the southwest over a bedrock outlet. During late summer and fall the outlet is dry. You'll see bedrock slabs and sand in the center of the outlet channel. Bedrock islands poke out of the shallow water near the shore. During spring, and commonly into midsummer, the trail is under water that can be chest deep. However, even when the water is high it doesn't flow particularly rapidly because the spillover fills a broad meadow to the southwest, making another shallow lake.

The inlet streams typically dry up by July or early August, depending on how much precipitation fell the previous year, and then the outlet stops flowing a few weeks later. This shutdown of the water supply occurs because the catchment area is largely bare, glacially scoured bedrock with little soil. Were the soils thicker, they would store more water and release it more slowly, extending stream flow later into the fall, and perhaps year-round. Once the outlet stops flowing, the lake loses water largely through evaporation. During the heat of the summer evaporation rates are estimated to be about 0.33 inch per day. The yearly range of water level in the lake has been measured via sensors that record water pressure, hence depth. The total range is typically about 5 to 6 feet; the

The outlet of Tenaya Lake during the high flow of spring (left), *in May 2007, and no flow in the fall* (right), *in October 2007.*

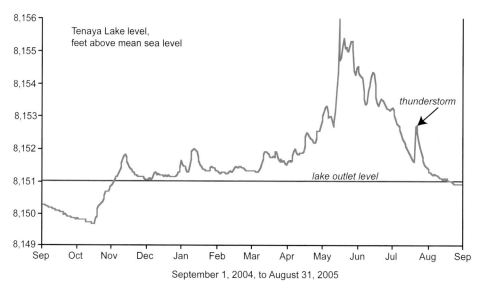

Variation in the surface elevation of Tenaya Lake from September 1, 2004, to August 31, 2005. The graph shows that there can be rapid, short-term fluctuations. Your authors were in the park during the July thunderstorms that produced one of the major spikes on the graph. Late in the afternoon of July 21, massive thunderstorms formed over the range crest near Mt. Lyell. Although these produced a spectacular orange sunset over the Valley, no rain fell there. But the next morning the Merced River was muddy brown, and its discharge had tripled overnight. A sheet flood came roaring off the bare rock of Stately Pleasure Dome and washed over Tioga Road at Tenaya Lake, and the lake's level rose 1 foot. As is evident from the graph, most of this excess water flowed out of the lake within about a week. —Data courtesy of Ned Andrews, U.S. Geological Survey

A glacial erratic along the outlet of Tenaya Lake with pollen "bathtub rings" that mark successively lower lake levels that occurred after spring runoff. The man is pointing at the lower limit of large greenish yellow lichen patches; this level may reflect the highest levels the lake reaches, for these lichens don't like to get wet. The photo was taken in October 2004.

lake drops 2 to 3 feet below the outlet in the fall and rises 2 to 3 feet above it during peak spring runoff.

A record of these fluctuations is partially recorded on the large boulders that line the outlet (which are glacial erratics, by the way). These typically show several prominent "bathtub rings" of pollen left behind as the lake level receded after spring runoff. Of particular interest are the large, bright greenish yellow lichen patches. Since these lichen patches live for hundreds of years and cannot survive prolonged submersion, the base of the lichen zone is probably a good marker for the high-water mark of the lake. Standing at this point, it is difficult to see how the lake could rise much higher, given how broad the outlet is and how much meadowy area there is to be flooded nearby.

Now that we know something about the hydrology of the lake and how much its level varies, let's see what it has done in the past. Proceed to stop 2. Gazing at the lake from its western shore, you will see something very odd: the gray, weathered top of a tree sticking a few feet out of the lake's surface. During peak runoff the lake might be high enough

View of one of the dead trees from a kayak. The tree is rooted in the lake bottom many feet below.

to cover this treetop, but most of the year it is exposed. Eight others are scattered around the lake, mostly along the far shore. They are lodgepole pines, the same as most of the trees around the lake. These aren't simply stumps sticking out of shallow water. Divers have found that their bases are apparently rooted in the rock on the lake bottom, 25 to 60 feet below the surface.

Lodgepole pines grow around the lake down to the typical high-water mark, but none seem to grow below that elevation. So why are there large trees sticking out of the lake? The most obvious answer is that at one time the lake was much smaller than it is now, or even completely dry. Lake level has not dropped more than 5 or 10 feet in historic time, so if such a drop happened, it must have been quite a drought.

Scott Stine, a professor at California State University, East Bay, collected wood from the outermost rings of two of these dead trees for radiocarbon dating. He obtained ages of around AD 1330 and 1090. These dates indicate that the trees died during the Middle Ages; the older age was the time of the Norman Conquest, and the younger age marked a time of crop failure, famine, and the Black Death in Europe.

By themselves, two curious radiocarbon dates on trees in the middle of a mountain lake might not mean much. But they meant a lot to Stine, because they corroborated a larger set of radiocarbon ages he obtained from stumps that had been submerged in Mono Lake. During the prolonged lake drawdown that occurred after Los Angeles began diverting water from streams that feed Mono Lake (see vignette 22), thousands of stumps of Jeffrey pines, cottonwoods, and various brushy plants were exposed. The inescapable conclusion is that the lake fell to that low level (without any help from Los Angeles) at some time in the past. The stumps occur in two different rings around the lake basin, each ring at a different elevation. The inner, lower ring of stumps yielded ages between AD 1000 and 1250, and the outer, higher ring yielded ages ranging between AD 1250 and 1400. The way they occur in rings, along with the different age ranges obtained through radiocarbon dating, indicates that the stumps record two different droughts. Each ring of stumps contains individuals with at least fifty growth rings, meaning that each drought lasted at least fifty years—they were megadroughts. The ages of the Tenaya trees fall within these age clusters, suggesting that they record the same two droughts. Similar age clusters are found in tree stumps rooted in the West Walker River near Chris Flat Campground, along US 395 about 3.5 miles north of Sonora Junction.

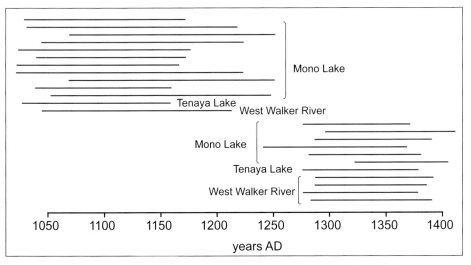

Radiocarbon ages of wood from trees in Mono Lake, Tenaya Lake, and the West Walker River. Each line represents the age of a single sample, with the approximate margin of error. Paleoclimatologists interpret these age clusters as representing the end of two droughts, when water flooded the area where the dead trees now stand. (Figure from Stine, 1994)

This general time of drought, AD 1000 to 1400, is further supported by tree-ring studies in the western United States. Tree ring widths vary with climate and can be compared to modern tree rings and records of temperature and precipitation to derive correlations. These relationships can be used on older wood to estimate ancient climatic conditions going back 1,000 years or more. Studies indicate that the West experienced an abnormally warm period between AD 1100 and 1375 and an epic drought between AD 900 and 1300, and the Colorado River basin experienced a severe decades-long drought around AD 1150. Although radiocarbon dates are not exact and times of drought may vary from place to place, it is clear that the first half of the second millennium was a dry time in the West.

These events roughly correspond to the Medieval Warm Period, a climatic warm swing that lasted about 400 to 500 years and preceded the Little Ice Age, a time of abnormally cold global climate between about AD 1350 and 1850. The Medieval Warm Period is also known as the Medieval Climate Optimum because in Europe, where it was first recognized, it led to warmer, locally more optimal conditions. In the western United States, though, it manifested as extreme drought, which isn't exactly optimal. This illustrates that whether a particular form of climate change is good

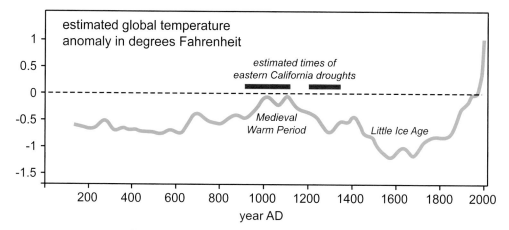

One estimate of global temperature variation over the last 2,000 years, reconstructed by Anders Moberg of Stockholm University and colleagues from a variety of sources, including tree rings and lake and ocean sediments. The temperatures today are significantly higher than those estimated for the Medieval Warm Period, and yet water persists in Tenaya Lake. This is a paleoclimatological mystery. Note that the second drought period in California does not correspond with the Medieval Warm Period, showing that local and global climate effects are not necessarily synchronized. The zero point for the temperature anomaly calculation is the global average of temperatures measured between 1961 and 1990.

or bad generally depends on location and perspective. In Europe, the Medieval Warm Period allowed grapes to be grown in the British Isles and permitted Vikings to establish colonies in Greenland. In western North America the record is less clear, but this period corresponds with the departure of the Ancient Pueblo Peoples from the Four Corners region, probably owing to drought-induced crop failures.

How bad were the droughts? Tree-ring studies in the Colorado River and Mono Lake basins indicate that the drought periods were characterized by 15 to 20 percent less precipitation than the average in the twentieth century. However, by balancing inflow, outflow, evaporation, and seepage from Tenaya Lake, Ned Andrews estimates that such a decrease would not have dropped the lake's levels far enough to allow trees to colonize lower elevations of its basin. He estimates that precipitation would have had to drop 80 to 90 percent lower than present values for trees to have grown where we see them in Tenaya Lake. This percentage seems remarkably high, and it is possible that something is missing from the analysis. Only further research will tell.

What causes precipitation and temperature to change for long periods? This question is at the heart of a great deal of current paleoclimatological research. Much of North America's weather variability is driven by changes in the position of the jet stream, the high-altitude band of strong winds that directs storms eastward across the continent. When the jet stream swings south, winter storms are directed into central California; when it swings north, they miss the state. Tree-ring studies of the megadrought suggest that the winter storm track stayed on a northerly track for a long time during the Middle Ages.

It is clear that many climate patterns are driven by changes in sea surface temperatures, and that warming or cooling in one part of the ocean can affect weather far away. For example, the well-known El Niño and La Niña events are changes in sea surface temperatures in the equatorial Pacific that affect amounts of precipitation in North and South America. Climate scientists have identified several other climate oscillations of this sort that operate on time scales of decades or longer. It is possible that changes in sea surface temperature caused by subtle changes in the sun's brightness and by volcanic eruptions, especially in equatorial regions such as Indonesia, play a role in long-term climate variations.

Regardless of the cause, evidence from Tenaya Lake and numerous other areas in North America shows that California has undergone massive droughts in the past, some lasting at least decades based on the life spans of trees that colonized formerly wet areas. Food for thought for inhabitants of the western United States: should another megadrought grip the West, the results would be catastrophic.

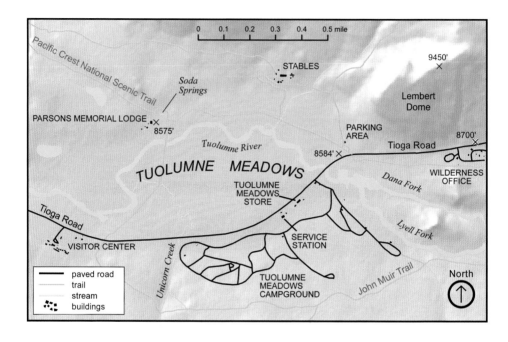

GETTING THERE

Soda Springs lies on the north side of the Tuolumne River. From Big Oak Flat Entrance Station, follow California 120 east (Big Oak Flat Road) for 7.8 miles. Turn left, continuing on California 120 east (Tioga Road) for approximately 38.6 miles. From Tioga Pass Entrance Station, follow California 120 west (Tioga Road) approximately 8 miles. Park at the trailhead (0.25 mile east of the turnoff to the Tuolumne Meadows Visitor Center and 0.75 mile west of the Tuolumne Meadows store) on the north side of the road. The route to Soda Springs is an easy, flat hike of about 0.5 mile. Look for trail signs to Soda Springs and Parsons Memorial Lodge. You can also access Soda Springs from the Lembert Dome parking area to the east. From that trailhead it is an equally gentle walk of less than 1 mile.

16

Soda Springs

THAT FIZZY TASTE CARRIES A GEOCHEMICAL SURPRISE

Hikers in the Tuolumne Meadows area will encounter signs to Soda Springs, a small, rust-stained spring that bubbles up inside a wooden enclosure just east of Parsons Memorial Lodge. The spring is thought to be the place where, beside a campfire in 1889, John Muir and Robert Underwood Johnson, an influential magazine editor, conceived of the idea of Yosemite National Park. The lodge was built in 1915 by the Sierra Club and now hosts interpretive talks. Many generations of hikers have drunk from the spring (although drinking this water is not advised owing to the low but ever-present threat of surface contamination), but few appreciate the geologic forces working beneath their feet to produce this unusual soda spring.

The word *soda* is used to denote many different things. To most Americans it probably connotes a beverage made fizzy with carbon dioxide gas. The term is used for any number of chemical compounds, including sodium carbonate, an innocuous chemical used to produce glass and soap, and sodium hydroxide, or lye, a terribly caustic compound used in many manufacturing processes and to unclog pipes. In the kitchen *soda* refers to sodium bicarbonate, otherwise known as *baking soda*. In the latter cases *soda* refers to the element sodium, but in drinking sodas and soda springs it refers to carbon dioxide bubbles. How these quite different chemical compounds came to be referred to by the same name is lost in the mysteries of language.

Although all natural waters have some carbon dioxide in them, and there are hundreds of natural springs in the Sierra Nevada, bubbling springs rich in carbon dioxide are not common in the Sierra Nevada; a compilation by USGS scientists lists only a dozen between Lake Tahoe and Sequoia National Park, many of which are named with some variation on Soda Spring. The two closest to this Soda Springs are near Mono Lake, one of which is warm (about 90 degrees Fahrenheit). To geochemists, bubbling springs are more than just a curiosity, as they contain information about the source of the water and nature of the rocks from which they issue.

Water in a spring can come from many sources, including local precipitation that soaks into the ground and reemerges rather promptly at the spring, groundwater that has spent a long time underground and may

Soda Springs in 1927, with sharp-tipped Unicorn Peak in the background above the hiker.
—Courtesy of the National Park Service

have traveled a great distance, or steam emitted from crystallizing magma deep below the surface. Similarly, carbon dioxide gas in springwater can come from many sources, including dissolved carbonate rocks (such as limestone), decomposing organic matter, and gases given off by magma. So where does the water in Soda Springs come from, and where does it get its bubbles? Let's start by figuring out where the water comes from.

Tracing a water's source is possible because both of its atomic constituents, hydrogen and oxygen, come in more than one isotopic form. What does that mean? Bear with us as we explain, because isotopic studies of hydrogen and oxygen play an important role in understanding the water cycle and past climates.

Most atoms of the element hydrogen consist of an electron orbiting a nucleus made of a single proton. The number of protons in the nucleus of an element is its atomic number, and the total number of protons and neutrons is its atomic weight. For normal hydrogen, both numbers are 1. But there is another form of hydrogen called *deuterium* (a mere 0.015 percent of natural hydrogen) that has a neutron in the nucleus, so its atomic weight is 2, although its atomic number is still 1. Deuterium, also known as *hydrogen-2*, is fully twice as massive as normal hydrogen, and this makes it behave differently than normal hydrogen. Alternate forms of an element with different numbers of neutrons are called *isotopes*.

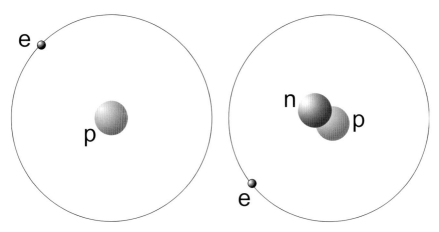

Schematic representation of atoms of hydrogen (left) *and hydrogen-2, or deuterium* (right), *with e = electron, p = proton, and n = neutron. Both have one proton in the nucleus and one electron. The extra weight of the neutron in hydrogen-2 makes water molecules containing hydrogen-2 behave slightly differently during chemical processes, such as evaporation, than water containing "plain" hydrogen.*

Oxygen comes in three different isotopic forms. The nucleus of every oxygen atom contains 8 protons. Oxygen with 8 neutrons (and thus an atomic weight of 16) makes up 99.76 percent of natural oxygen; oxygen with 10 neutrons (oxygen-18) makes up 0.20 percent; and oxygen with 9 neutrons (oxygen-17) makes up 0.04 percent. Oxygen-18 is 12.5 percent heavier than oxygen-16, so its behavior, too, is slightly different from its more abundant sibling. The differences in atomic weight among different isotopes of hydrogen and oxygen can be used to track where water comes from.

Ocean water, which is generally well mixed, has essentially the same ratios of these hydrogen and oxygen isotopes everywhere at any given time. However, when ocean water evaporates and forms clouds, the clouds tend to have slightly less oxygen-18 and hydrogen-2 than the ocean water left behind, because lighter molecules evaporate slightly more easily than heavier ones. So the next time you're at the beach, peer at the clouds and the water and then tell your friends, "Looks like those clouds have less oxygen-18 than the seawater." Chances are you're right, and they sure will be impressed (or at least amused).

What happens when those clouds make rain? The heavier isotopes condense a bit more easily than the lighter ones, so as the rain falls the clouds become even "lighter"—more enriched in the light isotopes of oxygen and hydrogen. Thus, the average isotopic composition of rain and snow in a given region is primarily a function of how much rain has been wrung out of the air masses that brought it to that area. In the continental

United States, the most wrung-out air masses—those with the least oxygen-18 and hydrogen-2—occur in the Great Basin and the Rocky Mountains, where mountain range after mountain range have sucked a great deal of moisture out of Pacific storms since their landfall. The Sierra Nevada, owing to its great height, has a large effect, and rain falling to the east of it is isotopically much lighter than that falling on its western slopes.

Determining past changes in the isotopic composition of seawater is one of the ways that geochemists learn about former climates. They don't work directly with ancient seawater, of course; they work with shells of marine animals. When an animal uses oxygen from seawater to build its shell, it picks up a record of the isotopic composition of seawater at that time. During times of widespread glaciation (such as 20,000 years ago, the time of the Tioga glaciation), a lot of isotopically "light" water (enriched in oxygen-16 over oxygen-18) is stored on land as ice. This leaves the oceans, and the animals that live in it, with a relative abundance of oxygen-18. Studies of the isotopic oxygen composition of ancient shells of a known age is one of the primary ways geologists have determined the timing of Earth's glacial cycles over the past several million years.

Now, given that lengthy introduction, what does this tell us about Soda Springs? First, let's look at the water temperature. The spring puts out water that is about 45 degrees Fahrenheit. That's chilly, but it's significantly warmer than the average annual temperature at Tuolumne Meadows. This means that the water isn't coming from the shallowest parts of the bedrock and soil, which would have temperatures similar to the surface, but from deeper bedrock. Temperature increases with depth at a rate of about 45 degrees Fahrenheit per mile in this region, so we can make a crude guess that, at a minimum, the water rises from roughly 1,000 feet below the surface.

The dissolved gases in Soda Springs are more than 99 percent carbon dioxide, with a little nitrogen and traces of argon and oxygen. Scientists have measured the isotopic compositions of the oxygen, hydrogen, and carbon in the Soda Springs water, along with the concentrations of several elements. The isotopic compositions of oxygen and hydrogen match local precipitation rather closely, so that appears to be what Soda Springs's water is: recycled rain and snow, rather than groundwater from far away or water that is the by-product of crystallizing magma. Local precipitation soaks into the ground, where it's mildly heated, and resurfaces at the spring. That isn't much of a surprise. However, the carbonation and isotopic composition of the carbon do come as a surprise.

There are two stable isotopes of carbon (stable meaning they don't morph into other isotopes): carbon-12, which accounts for 99.89 percent of natural carbon, and carbon-13, accounting for 1.11 percent. The carbon in Soda Springs water is moderately "light"—that is, enriched in carbon-12. This is characteristic of carbon derived from Earth's mantle, and of magmas derived from melting of the mantle. Isotopic studies

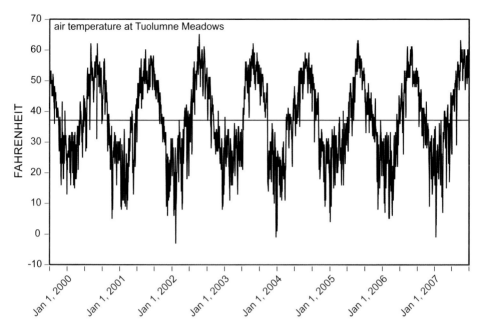

Daily average air temperature at Tuolumne Meadows. The yearly temperature swings are clear, and it looks like a few winters (such as the winter of 2002–2003) were a bit warmer than others. The average temperature (horizontal line) was 36.9 degrees Fahrenheit, fully 8 degrees cooler than the water issuing from Soda Springs.

show that it isn't extremely light, a characteristic of living organic matter, nor is it heavy (enriched in carbon-13), as are the shells of marine organisms and the rocks, such as limestone, formed from them. The water's isotopic carbon composition is similar to that of the carbon dioxide gas being released into the soil around Mammoth Mountain (see vignette 27 in *Geology Underfoot in Death Valley and Owens Valley*). The gas has killed thousands of trees in an area covering over 100 acres. Since carbon dioxide is food for trees, it may seem counterintuitive that it would kill them, but trees take in the gas through their leaves and needles, whereas they're poisoned if too much carbon dioxide surrounds their roots. The Mammoth Mountain area is volcanically active, so it is quite likely that the gas is being expelled from crystallizing magma deep below the surface.

Could the bubbles in Soda Springs really be derived from volcanic activity? It's possible. The youngest evidence of volcanism near Tuolumne Meadows is Little Devils Postpile, volcanic rocks that were intruded into the region about 8 million years ago (vignette 18). That isn't a reasonable source for the carbon dioxide in Soda Springs because volcanism has been dead there for millions of years. However, active volcanoes occur about 18 miles to the east in Mono Basin (see vignette 22) and to the south at Long Valley Caldera. Given this nearby volcanic activity and the abundance of fractures in the Sierra Nevada's rocks caused by faulting, it is possible that gases derived from crystallizing magma deep below the eastern Sierran front could make it to the surface at Soda Springs.

Does this mean that Tuolumne Meadows is ripe for an eruption? No. However, the same cannot be said for Mono Basin and Long Valley Caldera. Whatever their origin, the bubbles in Soda Springs water remind us that magma lurks just beneath the surface of Yosemite's backyard.

17

Runaway Rocks

METAMORPHIC ROCKS AT MAY LAKE

Bedrock within Yosemite National Park is nearly all granite, and it is this foundation that accounts for the beauty of the landscape. Metamorphic rocks occur outside the park entrances on the east and west (see vignettes 19 and 25, respectively) and produce a much different and darker landscape. However, a small body of diverse, colorful, and enigmatic metamorphic rocks occurs in the heart of the park at May Lake. This beautiful lake is also the jumping-off point for hiking up Mt. Hoffmann, a must-do hike for any serious Yosemite-phile.

Metamorphic rocks caught up in bodies of granite are prized by geologists because they offer clues about what kind of rock a region was composed of before the granite magma invaded it. The metamorphic rocks at May Lake tell an interesting story. They seem to have originated far to the south, in the vicinity of the Mojave Desert, 250 miles away. They are runaways.

Before beginning our traverse, let's review a few things. First, we remind you that collecting rocks and minerals in Yosemite National Park is strictly prohibited. Second, please be kind to the landscape. The May Lake area is heavily used, and much of the vegetation around the lake has been trampled. Please stay on existing trails or rock slabs. Finally, as you always should be while in Yosemite, be aware of bears. Be sure that your vehicle is free of food, smelly objects, or anything that looks like food or a smelly object.

The May Lake parking area is located on the old Tioga Road. Beyond the parking area the road is now a trail, and if you were to follow it you would end up about 0.5 mile west of Tenaya Lake. This section of road was replaced by the current stretch, to the south, in 1961. The parking area is located next to a chain of small ponds, and in early summer the mosquitoes can be fearsome. Rocks around the parking area are composed of Half Dome Granodiorite, and our hike will take us from this unit, through the outermost granodiorite of the Tuolumne Intrusive Suite (the granodiorite of Kuna Crest), to metamorphic rocks.

Hike north along the trail to May Lake, past outcrops of Half Dome Granodiorite on the left. These outcrops contain numerous dark enclaves, pieces of diorite that were molten when they mixed with the magma

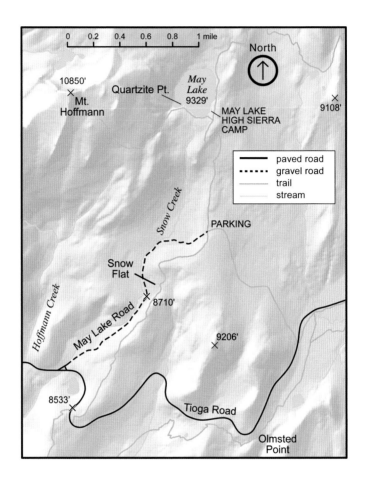

GETTING THERE

May Lake is located at the eastern base of Mt. Hoffmann, north of California 120 (Tioga Road), and is reached by an easy hike of about 1.2 miles from the May Lake Trailhead. From Big Oak Flat Entrance Station, follow California 120 east (Big Oak Flat Road) 7.8 miles. Turn left, continuing on California 120 east (Tioga Road) approximately 27 miles. From Tioga Pass Entrance Station, follow California 120 west (Tioga Road) approximately 19.5 miles. Turn north and drive 1.8 miles up the poorly paved road to the trailhead. Park at the trailhead, and be sure to take all food out of your car and leave it in the metal lockers to avoid bear break-ins.

of the Half Dome Granodiorite (see vignette 1). The trail proceeds only slightly uphill and almost due north, through the forest, for about 0.4 mile, and then starts a slightly steeper climb, swinging to the right and into the open before a sharp bend to the left (southwest) 0.3 mile ahead. Before the bend there are many more of these dark diorite blobs, but you will see pieces of other rock embedded in the granodiorite as you make the sharp left-hand bend. These other rocks, most of which are gray to rusty brown, are metamorphic rocks that were incorporated into the granodiorite magma as it intruded the crust. The magma melted some of the bedrock it intruded, but these pieces did not melt completely.

The proper term for a piece of foreign rock caught up in magma is *xenolith*, which, translated from its Greek roots, means, not surprisingly, "foreign rock." Although the diorite enclaves may seem as foreign to the host granite as these xenoliths, they are given their own name because they were molten, and thus part of the magma system, when they were introduced into the granite magma. Most of the xenoliths along this stretch of trail are grayish pieces of quartzite (metamorphosed sandstone) with obvious layering. They stand out well from the surrounding granodiorite and are commonly coated with a red iron stain that makes them even more evident. They occur as isolated pieces, typically fist sized, in great swarms. Hundreds to thousands are evident along the stretch of trail that heads southwest along the base of the steep slope that leads up to May Lake. Along this stretch of trail you pass from the Half Dome Granodiorite (on the east) into the granodiorite of Kuna Crest (to the west), but the contact is gradational and subtle.

After you encounter the xenoliths, the trail switches back right and left and then makes a sharp right-hand bend and ascends a steep slope up to the level of the lake. As you reach the flat area around the lake, the trail enters the forest. Bedrock is not exposed here because it is covered by till deposited by a Tioga-age glacier about 20,000 years ago (the till also forms a natural dam that increases the size of the lake), but many of the boulders lining the trail are samples of the local bedrock. A short while later you'll come to toilets, food storage lockers, and a faucet with drinking water at the southern end of the May Lake High Sierra Camp (the toilets and faucet only work during summer).

At this point the trail is on the contact between the granodiorite of Kuna Crest, on the east, and the metamorphic rocks around and under the lake, but the contact is covered by till. Walk past the water fountain to the edge of the lake and follow the trail to the west (left), toward Mt. Hoffmann. About 50 yards after crossing the outlet of the lake, the trail forks; stay on the main branch, to the left, and cross a low rise made of brown metamorphic rocks. There's a little bay on the right, and our destination is the stretch of white rocks on the far side of the bay. Watch for beautiful lavender and green metamorphic rocks on both sides of the trail. About 200 yards past the lake outlet, still on the trail to Mt. Hoffmann, you'll

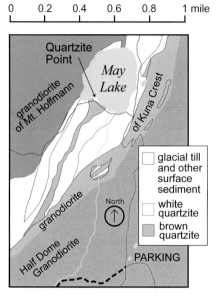

0   0.2   0.4   0.6   0.8   1 mile

Simplified geologic map of the area around May Lake. The lake sits in a bowl carved out of highly jointed metamorphic rocks, which are largely white quartzite and grayish brown quartzite. The contact between the Half Dome Granodiorite and the granodiorite of Kuna Crest is gradational. (Based on mapping by Ryan Taylor, MS thesis, University of North Carolina at Chapel Hill, 2004.)

come to the prominent band of white rocks that juts into the southwestern part of the lake. This site, which we call this Quartzite Point, is a good place to reconnoiter the landscape.

Looking toward the center of the lake, the rock that makes up the steep slopes to the left (west) is the granodiorite of Mt. Hoffmann, a widespread unit that greatly resembles the El Capitan Granite and is about the same age (see vignette 1). This unit touches the metamorphic rocks along a contact that runs north-northeast along the lake's western shore. The eastern contact of the metamorphic rocks, against the granodiorite of Kuna Crest, is near the eastern edge of the lake and under the High Sierra Camp. Before granite invaded, this area, including all the volume now occupied by granite, was composed of sedimentary rocks—sandstone, shale, limestone, and their kin—which were cooked by the granite into the metamorphic rocks seen here. Where did all those rocks go? Did the granite melt them? Push them aside? Push them up to be eroded? Geologists have debated these questions for over a century and continue to do so, with Yosemite as a centerpiece of the discussion.

It's no coincidence that the lake lies in the metamorphic rocks; the highly jointed and relatively weak metamorphic rocks were no match for the glacial ice that poured across this area on its way to Yosemite Valley, so the ice plucked out rocks and formed the lake's basin. The ice sculpted the white quartzite near the trail into a long, low, rounded form that looks like a giant baguette. Features like this, known as *rock drumlins*, are oriented parallel to the direction the glacier flowed, which was roughly north to south in this area.

The white quartzite is nearly 100 percent quartz. This rock was originally a beach sand, similar to the sand along many shorelines today, composed of quartz sand grains. This quartzite poses a problem to geologists trying to reconstruct the geologic history of the region. The sand was deposited between roughly 600 and 500 million years ago, and at that time this part of the world should have been at the bottom of a deep ocean, where only mud is deposited. Sand is deposited in nearshore environments—beaches—and rarely makes it into the deep ocean because ocean currents are too weak to carry the relatively large sand grains far from the edge of a continent.

When the rocks at May Lake were being deposited, the sandy shoreline of western North America ran southwest from Utah, through southern Nevada and western Arizona, to the Mojave Desert of southern California. We know this shoreline existed because geologists have mapped a belt of sandy rocks of similar age corresponding to the shoreline. For example, the Tapeats Sandstone of the Grand Canyon and the Zabriskie Quartzite of Death Valley were originally sands of this shoreline. Northwest of the sands, thick sequences of mud were deposited on a shallow continental shelf (similar to the continental shelf off the modern East Coast), today represented by the limestone and shale in, for example, the White Mountains of eastern California. And farther west still, in present-day western Nevada, the rocks are largely shales that formed from mud deposited in deep water.

Because of their geographic location, the rocks at May Lake should be part of this deepwater shale package, but instead they are sandstone, now metamorphosed into quartzite. A larger area of similar rocks occurs along the northern boundary of Yosemite National Park at Snow Lake, west of Tower Peak. These rocks, too, are far from the location of the ancient shoreline environment where they were deposited. How can this be? One explanation is the old shoreline had some sharp twists and turns, so the May Lake and Snow Lake rocks are where they were originally deposited. In 1990 Mary Lahren and Richard Schweickert of the University of Nevada at Reno proposed a different explanation: the Snow Lake rocks slid northward from the Mojave Desert to their present location along a major fault that ran through the Sierra Nevada. This explanation requires about 250 miles of movement—a large amount, but not an unusually large amount. (Rocks have moved a comparable distance along the San Andreas Fault.) They proposed that the rocks at May Lake were moved here along this fault.

Look closely at the quartzite and you will see that it is strongly deformed. There are many faults and folds, and in places you can find bits of darker quartzite that were torn into pieces as the white quartzite around them flowed and stretched during deformation at high temperatures. Given that these rocks are now sandwiched between the granodiorite of Mt. Hoffmann and the granodiorite of Kuna Crest, and that they

White quartzite with a grayish brown layer of mica-rich quartzite at Quartzite Point. The mica-rich layer was thinned and stretched by the white quartzite that flowed around it as the package of rocks was being deformed. Compass for scale.

The dark rock at Quartzite Point is a basaltic dike that intruded the white quartzite; it has been metamorphosed to blackish green hornblende. A similar dike near Snow Lake, at the north edge of the park, intruded similar quartzite and was dated at about 148 million years old. Similar dikes of the same age cut similar quartzites in the Mojave Desert, further cementing the tie between the rocks at May Lake and those in southern California. Knife for scale.

were likely moved here along a fault of San Andreas dimensions, it isn't surprising that they are so deformed. The rocks have taken quite a beating in the 500 million years they have been around.

This conundrum illustrates why metamorphic scraps like these at May Lake are so well-studied: they give important clues about paleogeography, the study of what Earth's surface looked like in ancient times. Paleogeographic studies are the means by which we know that Africa, Europe, and the Americas were joined together 200 million years ago, after which they rifted apart, forming the Atlantic Ocean and leaving beach sands along the present shorelines of those continents.

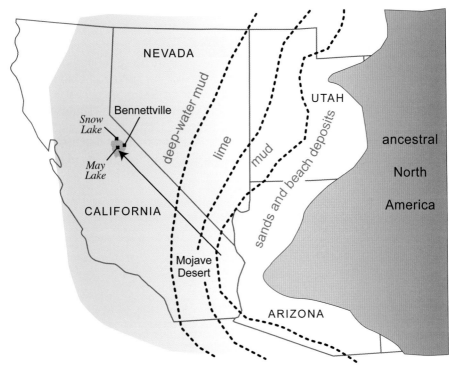

A schematic view of the West Coast of North America between 600 and 500 million years ago, when the rocks at May Lake were deposited in an ocean (blue), plotted on present-day state outlines. The shoreline moved back and forth as sea level rose and fell, sometimes reaching into southern Nevada and eastern California. The metamorphic rocks at May Lake seem to be grossly out of place since they formed from sand, a shallow-water deposit. The sand may have been deposited in the Mojave Desert region and then moved to its current location along a fault, the existence of which is only inferred (arrow). The metamorphic rocks at Bennettville (vignette 25) started as deepwater muds, consistent with their geographic position in relation to the ancient shoreline shown here. This evidence indicates that a fault must run between May Lake and Bennettville, since the locales are so close to one another yet the rocks of similar age that underlie them are so vastly different.

Many different varieties of beautiful metamorphic rocks occur on the south and east sides of the lake, all of which started as sediments deposited in the same nearshore environment as the sands of the quartzite. Most of these rocks came from a brownish sandstone similar to the sandstone that became the white quartzite, but containing a higher proportion of shale (originally mud). The shale formed the mineral biotite during metamorphism. Because biotite minerals grow parallel to one another in many metamorphic rocks, they impart a flakiness to the rock; this flaky rock is called *schist*. In places this brown sandstone underwent further metamorphism that caused dark and light minerals to segregate themselves into layers, forming the rock known as *gneiss*. Metamorphic rocks south of the lake include coarse-grained white marble and green-and-pink rocks that formed when the granodiorite magma reacted with limestone. These colorful rocks, known as *calc-silicate rocks*, are rich in the green mineral diopside and the pink to purple mineral garnet.

The southwestern shore of May Lake is the starting point for hiking up Mt. Hoffmann, named for Charles Frederick Hoffmann, a German-born topographer who was part of the 1860s geological survey of California. This is a strenuous but rewarding hike with an elevation gain of 1,500 feet from the lake and 2,000 feet from the parking area. The summit lies on the southern edge of a dramatic north-facing cirque that drops steeply about 1,000 feet to an unnamed lake. John Muir, in discussing how best to spend one's time in Yosemite, directed visitors to "go straight to Mount Hoffmann . . . From the summit nearly all the Yosemite Park is displayed like a map." We agree.

Root of an Ancient Volcano

Although the area east of Yosemite National Park is famous for its active volcanism (see vignette 22), volcanic rocks less than 150 million years old are rare within the park. There are several small outcrops in the remote southeastern section of the park and some near Hetch Hetchy Reservoir, but few visitors see these. However, there is a beautiful outcrop of volcanic rocks along the Tuolumne River downstream of Tuolumne Meadows and about 1 mile upstream of Glen Aulin High Sierra Camp. Most hikers walk right past this outcrop without seeing it.

This black outcrop, known as Little Devils Postpile, bears a striking resemblance to its well-known namesake, Devils Postpile near Mammoth Lakes. For some reason the devil gets credit for owning a lot of odd-looking real estate in the western United States, including Devils Golf Course, the jagged salt pan at the floor of Death Valley; Devils Cornfield, an area with oddly shaped bushes in Death Valley; Devils Tower, a volcanic neck in Wyoming; Devils Backbone, a prominent dike within the caldera at Crater Lake; and Devils Punchbowl, a depression or low area (six in California alone). There are 220 official place-names in California attributed to the devil, including canyons, corrals, gardens, gulches, holes, kitchens, slides, and thumbs.

Both Little Devils Postpile and its larger cousin display outstanding columnar jointing, contraction cracks that form in lava flows and other varieties of magma that cool rapidly. In rock bodies with well-developed columnar jointing, such cracks break the rock into roughly hexagonal columns whose long axes are parallel to the direction heat was flowing out of the magma as it cooled.

The Tuolumne River drains the northern half of Yosemite National Park before being captured in Hetch Hetchy Reservoir. Through much of its upper reaches it meanders lazily across hard granite bedrock, but once west of Tuolumne Meadows the river cuts through a westward-deepening gorge that is impressive enough to be called the Grand Canyon of the Tuolumne. This is the domain of several waterfalls, including Tuolumne, White Cascade, California, Le Conte, and Waterwheel. Little Devils Postpile crops out just upstream of where the deepening begins.

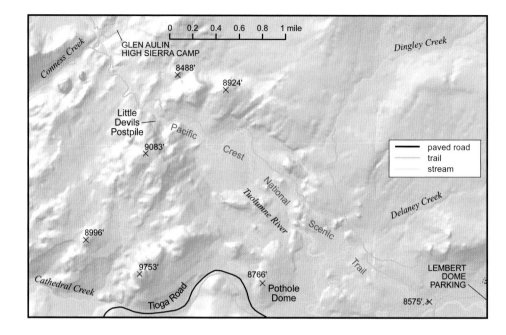

	paved road
	trail
	stream

Map labels: Conness Creek, Dingley Creek, GLEN AULIN HIGH SIERRA CAMP, 8488', 8924', Little Devils Postpile, Pacific, 9083', Crest, National, Tuolumne River, Scenic, Delaney Creek, 8996', 9753', 8766', Pothole Dome, Trail, LEMBERT DOME PARKING, 8575', Cathedral Creek, Tioga Road

Scale: 0 0.2 0.4 0.6 0.8 1 mile

GETTING THERE

Little Devils Postpile is located along the south side of the Tuolumne River, a pleasant and easy downstream hike of a little over 4 miles (one way) from Tuolumne Meadows. It is an excellent day hike with lots of beautiful scenery and rushing water. Start at the Lembert Dome parking area. From Big Oak Flat Entrance Station, follow California 120 east (Big Oak Flat Road) 7.8 miles. Turn left, continuing on California 120 east (Tioga Road) approximately 39.6 miles. From Tioga Pass Entrance Station, follow California 120 west (Tioga Road) approximately 6.9 miles to the Lembert Dome parking area. From the parking area, proceed west along the gravel road and trail, passing Soda Springs and Parsons Memorial Lodge and following the trail toward Glen Aulin High Sierra Camp. About 1.5 miles from the parking area a side trail from the stables joins the main trail. About 3 miles from the parking area the trail emerges from the trees to join the Tuolumne River, with a fine view of the Cathedral Range to the south across the meadow. Shortly after rounding a sharp bend in the river, the trail enters dense forest for about 0.25 mile. It then emerges from the forest and crosses bare slabs for a few hundred yards, goes back into the forest, emerges and crosses a second set of bare slabs, and goes back into the forest once again for a few hundred yards before leaving the forest for a bare rock ridge. You'll know you've reached the rock ridge when the trail rises steeply for a short distance and you have a rock wall on your right. The rocks here are full of huge orthoclase crystals, some up to 5 inches long. Little Devils Postpile is the prominent black outcrop across the river.

Before crossing the river for a closer look at the postpile, examine it from the rock slabs between the trail and the river. The river has carved a narrow gorge between the granite and the dark rock. You can walk right up to the edge of the gorge, which drops about 20 feet to the river, but the rock slants toward the precipice, so it is safer to examine the rock from some distance. Rock on the north (trail) side of the river is all Cathedral Peak Granite with impressively large and abundant orthoclase crystals. One can spend hours (or, in the case of your authors, many days) wandering around on these slabs examining the strange and wonderful sights, but that is not our goal today—our sights are set on the dark rock.

What is the dark rock? It looks a lot like basalt, the most widespread dark lava on Earth. However, chemical analysis shows that the rock of Little Devils Postpile is slightly on the weird side. It goes by the somewhat obscure name of *trachyandesite*, but we will call it *andesite* for simplicity's sake. The rock is similar to basalt but richer in the elements silicon, potassium, and sodium. Andesite makes up many of the big volcanoes in the chain of volcanoes that compose the Pacific Northwest's Cascade Range, such as Mount St. Helens.

Little Devils Postpile, looking south across the Tuolumne River. Columnar jointing is evident throughout the outcrop. In this view the columns mostly lean to the right.

From the trail the columnar jointing is well displayed. The columns appear to be leaning to the right. Closer to the river the columns look different, somewhat like a stack of sandbags. From that vantage you're seeing the ends of columns that are stacked horizontally. Somewhere in the flat ground just behind the riverside exposure, the columns bend to steeper attitudes. The contact between the andesite and the granite lies under the river. Even in the fall, when river flow is lowest, only granite is exposed on the north side and only andesite on the south. The attitude of the columns in this area provides an interesting clue to the origin of Little Devils Postpile, and we'll return to this later.

Although the postpile makes a fine sight from this side of the river, it is worth a closer look. Follow the trail about 500 yards downstream to the bridge, cross the bridge, and then backtrack along the river about 300 yards to the first outcrops of the postpile. Do not attempt to cross the river at the postpile, especially during high water! Once at the postpile it is easy to climb on top for a better view of the rock and columns.

The rock is fine grained like a lava, and few crystals are visible. Because lavas cool rather quickly on the surface, the crystals in the lava don't have enough time to grow very large. In contrast, the very large ortho-clase crystals in the granite across the river had ample time to grow as the magma cooled slowly beneath the surface. But if the postpile is a lava, then its location doesn't make sense: it lies in the bottom of the Tuolumne River canyon, right where large glaciers flowed. At this place the ice should have been more than 1,000 feet thick about 20,000 years ago and should have eroded away any lava flows. If the postpile was less than 20,000 years old, there would be no problem, but it is considerably older. Recent dating by your authors and Dr. Daniel Stockli of the University of Kansas indicates it is about 8 million years old.

There are two clues that tell us this outcrop is not a piece of a lava flow, as its fine-grained nature would lead us to believe, but is instead an igneous rock that cooled not far below Earth's surface. The first clue is the horizontal orientation of the columns along the river gorge. Columns in columnar-jointed rocks generally form parallel to the direction that heat flowed as it was conducted from the hot rock into cooler surrounding rocks or air. Joints in basaltic lava flows are generally vertical, perpendicular to the lava flow's top and bottom surfaces. Here, though, they are almost horizontal, indicating that heat was being conducted horizontally into the granite. This indicates that the contact between the andesitic magma and the granite was vertical. This is what one would expect in a dike, a steep, tapered sheet of igneous rock that forms when magma rises in cracks in solid rock and solidifies. However, the orientation of the columns is not undeniable proof that we're looking at an intrusive rock rather than a lava, because columns can form in such orientations in other settings, such as where lava flows along the side of a glacier.

View of Little Devils Postpile from the granite slabs north of the Tuolumne River. The rock looks like stacked sandbags; these are the ends of columns.

In this view looking upstream in the gorge during low water in October, granite lies on the left, and andesite on the right. The beach in the foreground is under churning water during much of the year. Because the granite is strong and relatively unjointed and the andesite is highly jointed, the river can more easily carve out and carry away the andesite. The vertical granite wall along the river is probably very close to the original andesite-granite contact.

Much better evidence is in the granite exposed on the uphill side of the andesite. The contact between the andesite and granite is well hidden by vegetation and soil, but in one spot unusual dark gray granite shows up very close to andesite. Closer inspection shows that this rock consists of rounded chunks of quartz and feldspars in a dark, almost black matrix. The dark material is glass, and the rock chunks are granite that was melted by the hot, intruding andesite. This composition indicates that, along the contact, the magma melted the surrounding granite by about 50 percent.

Now lava is hot, so one might ask why the granite couldn't have been melted by a lava flow pouring over it at Earth's surface. There are two reasons. First, if you put two materials with different temperatures in contact and let heat conduction take over, the temperature at the contact will be the average of the two temperatures. For example, if a lava flow at 2,200 degrees Fahrenheit (on the high end for lava of the postpile's composition) flows over a 50-degree-Fahrenheit rock surface, the surface of the rock will heat up to about 1,125 degrees Fahrenheit and then cool down as the lava cools. The melting temperature of granite at the surface is around 1,600 degrees Fahrenheit, so this average temperature wouldn't have been hot enough to melt the granite at all, much less melt 50 percent of its mineral constituents. The other reason we know this melting wasn't caused by a lava flow comes from analysis of rock under other lava flows around the world. Rocks covered by lava typically are baked to a red color but don't melt much, if at all.

Given the temperature argument, how could the granite have melted in the first place? If magma flowed through a dike adjacent to the granite for a long time, then it could have heated the granite nearly to the magma's temperature. The first magma through the crack would have chilled to the average temperature of the magma and host rock, but continued magma flow would have heated the interface more and more, eventually melting the host rock. Melted host rock is seen adjacent to many eroded dikes, helping solidify this nifty explanation for what we see at Little Devils Postpile.

From the dike explanation we can infer that there was an andesite volcano on the surface 8 million years ago, fed by the magma now solidified at Little Devils Postpile. How deep was the rock we're standing on at that time, when it was magma? Given that the andesite is fine grained like a lava flow rather than coarse grained like an intrusive igneous rock, the answer is probably "not very," but we have little else to go on. It could have been as shallow as a few hundred yards, or maybe 500 yards, or . . . we don't know. Answering this question is central to understanding how much rock subsequent glaciers removed from this area—in other words, allowing us to know the height of the volcano that existed here. What the landscape was like during the early periods

This close-up of the melted granite adjacent to the postpile shows rounded blobs of feldspars (plagioclase and orthoclase) and quartz in a matrix of dark glass. Knife for scale.

of glaciation is also a mystery. Was there a Tuolumne River? Were the mountains high or low? Such questions are the focus of much current research.

Is it just coincidence that the river excavated a gorge right along the granite-andesite contact? Perhaps, but more likely the river followed the andesite down as it eroded its channel. If the dike we're looking at extended all the way to the surface, at some time in the distant past the ancestral river could have found this conduit and exploited its easily eroded, highly jointed rocks. The contact would have been a particularly good place to attack the rocks because of the extreme thermal stresses that occurred there.

Before leaving the area, you might want to walk around on the rock slabs between the trail and the river, where there are some truly strange and unusual sights. The orthoclase crystals in the Cathedral Peak Granite are particularly large here, some reaching 4 or 5 inches in length. In places, especially where the trail descends the downstream side of the small climb noted in the "Getting There" directions, orthoclase crystals make up 80 percent or more of the rock and form beautiful mosaics where the rock was glacially polished. Although some geologists have interpreted these mosaics as being piles of crystals that settled out of

A mosaic of orthoclase crystals on glacially polished granite slabs near the trail. These accumulations occur as irregular blobs or steep tabular bodies several feet thick. The compass is 3.5 inches long.

A ladder dike near the Tuolumne River, composed of layers of biotite and other dark minerals and abundant, large orthoclase crystals. The compass is 3.5 inches long.

mushy magma to the bottom of a magma chamber, they cannot be; they are packed too tightly to be physical accumulations; plus, orthoclase is the last mineral to crystallize from a magma of this composition, so the magma was mostly solid when these crystals were growing. The origin of these mosaics is currently a mystery.

Even stranger are the bizarre ladder dikes found on the granite slabs just upstream of the small climb, on both sides of the river. These strange, curving layers of biotite and other dark minerals, typically studded with large orthoclase crystals, occur in a band that runs from this area south toward Tioga Road near Fairview Dome. How these form is another mystery, one that has preoccupied geologists around the world for decades.

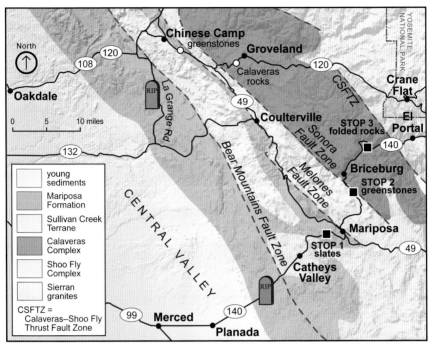

Simplified geologic map of the western approaches to Yosemite National Park. The different colors mark the different packages of rocks that have been juxtaposed by various faults, and yellow indicates recent sedimentary fill in the Central Valley. The stops on California 140 are marked, the white circles along California 120 mark alternative sites, and tombstones mark particularly good areas to see tombstone rocks.

GETTING THERE

Interesting metamorphic rocks occur along all the roads leading to Yosemite from the west, although the road in from the south, California 41 from Fresno, lies almost entirely in granite and related rocks of the Sierra Nevada Batholith. This vignette makes several stops along California 140, but you can see many of the same rocks along California 120 and California 132.

You can see "tombstone rocks" along California 140 between Merced and Catheys Valley. To see slates, greenstones, and folded rocks, drive east from Catheys Valley, resetting your odometer at the county park in Catheys Valley. Drive 6 miles east to Dial's Rock Shop and park a few tenths of a mile beyond the shop, on the right; this is stop 1. On the way to stop 2, California 140 passes through Mariposa and Midpines. Note your mileage at the county park in Midpines, then continue 3.1 miles to a large roadcut about 0.5 mile past the point where the road starts to head downhill; park on the left, at stop 2. After leaving stop 2, the highway drops into the deeply incised canyon of Bear Creek. At Briceburg the road begins to follow the Merced River canyon. Although it's difficult to know one's orientation in the deep, meandering canyon, the road proceeds northeast from Briceburg for 4.5 miles before bending due north to skirt the north end of Ferguson Ridge. The road and river go north for 1 mile, east for 1 mile, and south for 1 mile; shortly after the bend to the south there is a geological exhibit, marked by a brown sign proclaiming the "Oldest Rocks of Yosemite Region"; this is stop 3. Park in the large pullout and carefully cross the highway to the river.

19

Tombstone Rocks, Slate, and Greenstone
ROCKS OF THE WESTERN APPROACHES

Yosemite is carved in granite, and it is no coincidence that the park's boundaries generally lie where granite gives way to other, more easily eroded rocks that rarely form such spectacular scenery. But these other rocks have charms of their own. In this vignette we describe some of the older rocks found along the western approaches to the park and detail what they tell us about the geologic history of the area.

The rocks that predate the granite in the Yosemite region are mostly between about 600 and 150 million years old, significantly older than the 105- to 85-million-year-old granites that make up most of the park. They are complicated, messy, and covered with vegetation, but they have received a great deal of geologic attention for three reasons. First, they contain information about the geologic history of central California during that period—just about the only record we have. Second, the tectonic deformation that the rocks have suffered tells us about plate collisions that have occurred along the western margin of North America. Third, they tell us what rocks existed in the area of the present Sierra Nevada before the granites arrived.

This last point bears elaboration. If you stand on a high point in Yosemite National Park, virtually every nonliving thing that you can see is granite. Geophysical studies tell us that the granite is 6 miles or more thick, so there is a huge volume of granite, all of which was intruded into Earth miles below the surface. The granitic magma didn't fill holes in the crust, so whatever rock used to be in that volume is now gone. Where did it go? Was it pushed up above the magma and subsequently eroded away? Was it pushed below the magma? Did it dissolve in the magma? This topic, known as the "room problem," has occupied geologists for over a century, and the debate continues. The rocks west of Yosemite can help us answer these questions.

The rocks that predate the granite are all metamorphic rocks, which means they have been changed—metamorphosed—from their original nature. The original rocks in most cases were sedimentary and volcanic. Metamorphism occurs when rocks are subjected to heat (from burial or nearby injections of magma), pressure (burial), directed stress (from plate collision), or combinations of all three. These processes transform

205

one type of rock into another, but in the case of most of the rocks west of the park, the original rock type is still discernable.

Consider mud. If mud is buried by more mud and dries out, it becomes shale, a familiar rock. Shale consists of fine particles of quartz and various clay minerals. If the shale is buried more deeply and heated a bit, it will be metamorphosed into slate, a harder version of shale in which some of the clay minerals are transformed into aligned crystals of the mineral mica. This alignment causes the rock to break easily. Old blackboards were made of slate because the best slate breaks into nice planar sheets, is dark colored, and has plenty of hard, fine quartz crystals that grab chalk particles as the chalk slides by, leaving a mark.

Slate subjected to even higher temperatures and pressures can metamorphose into schist, a rock characterized by more abundant aligned mica crystals. Schist breaks easily along the plane of aligned micas, forming platy outcrops. The "tombstone rocks" of the foothills show this feature well.

Metamorphic rocks are typically layered. One might quite reasonably interpret such layering as original sedimentary layering, which develops as sediments are deposited by water and wind. In some cases it is, but layering can also develop during metamorphism. It is quite common for the plane of mica alignment in schists to be at a different angle than the original sedimentary layering.

Rocks of the foothills, west of the granites in the central and southern Sierra Nevada, occur in long, thin belts that parallel the mountain range; collectively, they are known as the Western Metamorphic Belt. In most cases these belts are bordered by fault zones, and the rocks on either side of the fault zones don't match up. That means there has been move-

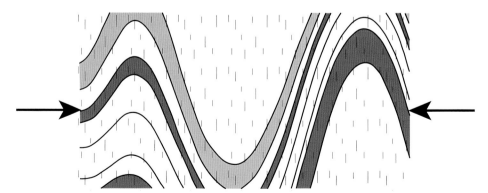

When shale is squeezed (arrows) and folded during metamorphism, the alignment of micas (short vertical dashes) can be at a high angle to the original sedimentary layers (wavy lines). If the metamorphism is intense enough, the layering produced by mica alignment can be the most prominent layering in the rock. Such layering is termed foliation.

ment along the fault zones, but for most of them, the direction of motion isn't well understood. Individual faults may be similar to the San Andreas Fault, where one body of rock slides horizontally past another; similar to the eastern boundary of the Sierra Nevada, where one rock body drops relative to the body on the other side of the fault; or similar to the faults on the south side of the San Gabriel Mountains in southern California, where one rock body is thrust up and over another. Generally, the rocks on either side of these faults are so different that they cannot be matched up across the fault to see how much displacement occurred as a result of movement along the faults. Geologists have a special word for such fault-bounded slices of rocks: *terrane*—a geological variation on *terrain* that refers to a fault-bounded package of rocks that differs from adjacent packages and may have come from some distance away.

The rocks, faults, and terranes of the foothills have been anointed with a perplexing variety of names. Although many are euphonious and exotic (Bear Mountains Fault Zone, Shoo Fly Complex, ophiolite of Devils Gate, and many others), we will keep the nomenclature simple and not try to sort out the various interpretations of these rocks and faults at the risk of offending geologists who have spent much of their adult lives trying to do so. All of these rocks and faults disappear to the west under younger sedimentary rocks that have filled in the Central Valley.

The westernmost rocks in the Western Metamorphic Belt belong to the Mariposa Formation. These are largely metamorphosed volcanic and sedimentary rocks 160 to 155 million years old, named for exposures around the town of Mariposa. The Mariposa Formation holds a special honor in California's geologic heritage because it hosts many of the gold-bearing quartz veins that led to the gold rush of 1849. These quartz veins, several of which are visible in the foothills along California 140, were produced by hot fluids that circulated through the metamorphic rocks the Sierran granites were intruding.

Driving northeast from Merced, you will notice small fins of rock sticking up out of the fields along the highway. These become especially noticeable about 8 miles northeast of Planada, shortly after the flatness of the Central Valley gives way to the foothills. These fins, known colloquially as "tombstone rocks," are thin layers of schist that have weathered into these shapes. (Excellent examples occur in many places along the foothills, including along La Grange Road, which runs south from California 120 west of Chinese Camp, and around Copperopolis.) Numerous white quartz veins stick out of the ground on the north side of the road among the tombstone rocks northeast of Planada. On the way to stop 1 you may notice that you pass into granitic rocks about 2 miles east of Catheys Valley. This is the Guadalupe pluton, which intruded the slates and schists.

Much of the Mariposa Formation consists of slate, and there are excellent exposures of these slates at stop 1. The massive outcrops across the

"Tombstone rocks" of the Mariposa Formation, together with an oak tree, are a common sight in the Sierra Nevada foothills.

highway from the rock shop are beautiful gray slates that weather to a rusty brown. They look like layered sedimentary rocks—shales, in this case—but they are not. They are metamorphic, having been dramatically folded and deformed, as have most of the other rocks along the west side of the park. Although the sedimentary layers look undeformed, they have been squeezed and tightly folded like a stack of fanfold printer paper. At stop 1 the layers dip steeply to the north, although in much of the foothills they dip to the northeast. Imagine a stack of fanfold paper tilted so that the layers are steeply inclined to the north, and you can visualize what you're seeing at stop 1.

The original sedimentary rocks were deposited in fairly deep water around an island arc that lay off the west coast of California a few hundred million years ago, and some of the sandy layers in the slate show evidence for grading, wherein the larger sand grains settled more rapidly than the smaller grains, producing a layer that grades from larger particles at the base to finer particles at the top. If you get close to the outcrop and look for graded bedding, you will probably find layers that give contradictory indications of which way was up when the sediments were deposited. This is one line of evidence that indicates the rocks were tightly folded.

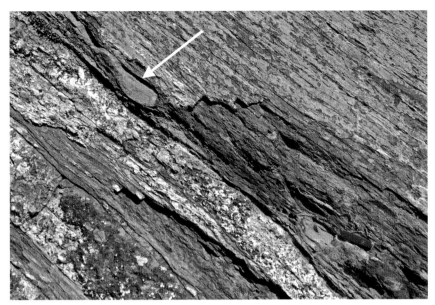

Although the Mariposa Formation slates may resemble nicely bedded, undeformed sedimentary rocks, they have been intensely squeezed and folded. One clue in this outcrop is how some layers look pinched and swollen. This feature is known as a boudinage, *from the French word* boudin *for "sausage," because beds that have been deformed in this way sometimes look like a string of sausages. A greenish metasandstone (metamorphosed sandstone) layer, just left of and under the knife in this outcrop, was pinched apart so that it is no longer attached to the rest of the layer (arrow).*

The slates at stop 1, in a roadcut along California 140 between Merced and Mariposa.

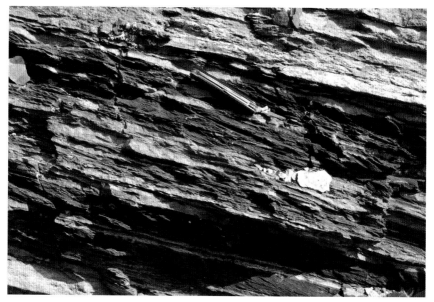

The main color variations in this outcrop of Mariposa Formation slate represent different sedimentary layers, which are inclined slightly. There is a noticeable set of closely spaced fractures running parallel to the knife. These fractures are parallel to aligned mica crystals, which grew during metamorphism and indicate that the rock was squeezed perpendicularly to the knife.

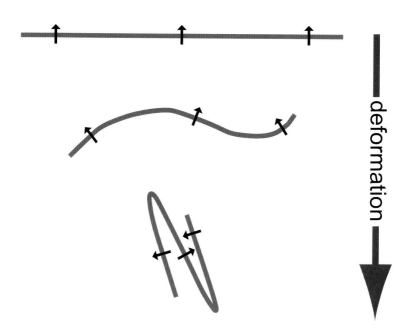

Folding can produce layering in rocks in which the original up-direction has been flipped back and forth. The arrows indicate the original up-direction when the brown layer was deposited; progressive folding produces a confused package of rocks with alternating up-directions. This is what is seen in the Mariposa slates.

As you make your way to stop 2, California 140 traverses the Melones Fault Zone just before entering Mariposa, crossing onto different rocks, known as the Sullivan Creek Terrane. The Sullivan Creek Terrane consists of volcanic rocks that are similar to those found in an island arc, such as the Japanese islands, and serpentine-rich rocks that formed when dark mantle rocks were metamorphosed. An island arc is a band of volcanic islands that developed over a subduction zone. Serpentine is a green, slippery-looking mineral rich in iron and magnesium. Serpentinite, a rock composed mostly of serpentine, is the state rock of California. California, first in so many things, was also the first state to have an official state rock and state mineral (gold). Some states don't even have an official state mineral or rock—sad indeed.

To a geologist, this assemblage of rocks (island-arc volcanic rocks and metamorphosed mantle rocks) buried within a continent raises eyebrows because it suggests that an island arc smashed into the edge of the continent in the past. One way this can happen is if an island arc rides along on a plate that is being subducted and makes it to the subduction zone. Island arcs are too thick and buoyant to be subducted (as are continents), so they jam up the subduction zone and get stuck on the edge of the continent rather than traveling into Earth with the rest of the plate. Geologists use phrases such as "the island arc slammed into the continent," but, of course, this process takes place over geologic time, with the slamming happening at a rate of perhaps 1 inch a year. Collisions of this sort probably produced much of the deformation we see in the foothills rocks.

There is a large quartz vein exposed in a roadcut 0.5 mile north of where California 140 and California 49 split. The green rocks in the roadcut on both sides of the road are the serpentine-rich mantle rocks that were metamorphosed. At stop 2 the rocks are conspicuously green, although they weather to rusty brown. The green color is caused by green minerals, such as chlorite, epidote, and serpentine, all of which formed when the volcanic rocks were metamorphosed. These rocks are known as *greenstones*. Of particular interest are the elongate blobby structures toward the downhill end of the outcrop that have been interpreted as pillow lavas—lavas that were erupted underwater. When lava erupts underwater, it quickly chills and forms a crust, but magma pressure repeatedly breaks the crust, causing blobs of new magma to ooze out. These also crust over, and the blobs form piles on the seafloor that look like mounds of pillows. Pillow lavas are a clear indication of an underwater eruption and are further evidence that we are looking at island-arc rocks.

Another interesting place to view these greenstones is 1.5 miles farther north, where the road begins a steep descent toward the Merced River. California 140 follows the canyon of Bear Creek, which originates on the ancient erosion surface around Midpines and here plunges over a knickpoint and down into the more recently incised Merced River canyon (see vignette 21 for more on erosion surfaces and recent river incision).

Greenstones at stop 2. The large vertical oval region just left of center, about 3 feet long, has been interpreted as a lava pillow. If true, then these lavas were erupted underwater.

A parking area on the right (east) side of the road gives access to beautifully water-polished greenstones along Bear Creek. (Sullivan Creek Terrane rocks are also well exposed along California 120 between Chinese Camp and Don Pedro Reservoir.)

As the road descends into the canyon, it crosses another major terrane-bounding fault zone, the Sonora Fault Zone, and passes onto another terrane—the Calaveras Complex. Calaveras rocks are largely well-layered, oceanic metasedimentary (metamorphic sedimentary) rocks that are brown and gray in outcrops and compose large cliffs along this stretch of road, especially about 1 mile south of Briceburg. Calaveras rocks are exposed along the road much of the way to El Portal (and also in roadcuts along California 120 west of Groveland).

Even though the sign at stop 3 claims that you are viewing the "Oldest Rocks of the Yosemite Region," this isn't true. Many local metamorphic rocks are older, including those at May Lake (vignette 17). Park in the large pullout and carefully cross the highway to the river. Except during times of high water, beautiful, intensely folded metasedimentary rocks of the Calaveras Complex crop out in the river and on the eastern shore. These rocks are especially photogenic because the river has polished them. The deformed rocks are several hundred million years old, and many events probably contributed to the mess you see—island-arc collision, movement along the Sonora Fault Zone, and intrusion of the Sierra Nevada Batholith among them. The batholith shoved aside the metasedimentary rocks and sat upon them in its attempts to solve the "room problem," heating and squeezing the rocks as it did so.

About 0.5 mile east of the geological exhibit sign, nature recently created a new geological exhibit. Named the Ferguson rockslide, this slide became active in spring 2006, beginning with just a few small rocks dropped onto the highway on April 25. However, the slide quickly reached massive proportions and a 200-yard stretch of California 140 was buried by tens of feet of rock debris, closing one of the major routes into Yosemite National Park during the peak tourist season. The moving mass had a volume of about 800,000 cubic yards, enough that it would have taken about eighty thousand dump truck loads to haul it away. The top of the rockslide is marked by a large scarp, a cliff along which the rock mass slid. The rockslide of 2006 increased the height of the scarp by about 30 feet, but that's a relatively small proportion of the overall scarp height. This suggests that the 2006 event was a reactivation of an earlier rockslide

Folded fine-grained metasedimentary rocks of the Calaveras Complex exposed in the Merced River.

surface; cosmogenic exposure dating of the upper, older scarp by your second author suggests that the earlier slide occurred about 3,000 years ago.

It isn't entirely clear what caused the rock mass to move again, but it probably had to do with unusually high precipitation in the spring of 2006. Water lubricates slide surfaces and also adds weight to the sliding material. As of this writing the rockslide has stabilized, but it still moves slightly during intense rainfall. Although unlikely, it's possible that rapid failure of the entire rockslide could dam the Merced River, temporarily

The Ferguson rockslide in summer 2006. The white lines just above the river are barricades placed along California 140 before it was completely buried by rock debris. At the top of the rockslide mass, note the curving upper scarp that forms the headwall of the slide. The fresh, unvegetated lower part of the scarp was exposed by the activity in 2006. The weathered and vegetated state of the upper scarp suggests that the slide of 2006 was a reactivation of an ancient slide surface.

creating a reservoir. The U.S. Geological Survey installed a sophisticated monitoring system to detect movement in the rockslide area, motion from regional earthquakes, and sudden changes in the level of the Merced River, any of which might indicate massive failure of the slope.

As of this writing, traffic on California 140 was being diverted across the river on temporary bridges to get around the slide. Driving on this diversion, first east across the river, then back west to the original side, emphasizes the particular danger that steeply jointed rocks can pose to roadways. When driving back to the west side, it's clear that joint surfaces in the schists on the west side of the road dip steeply toward the roadway, at approximately the same angle as the hillslope, providing slide surfaces for rock slabs. As long as California 140 runs up the canyon of the Merced, road crews will have to deal with falling metamorphic rocks.

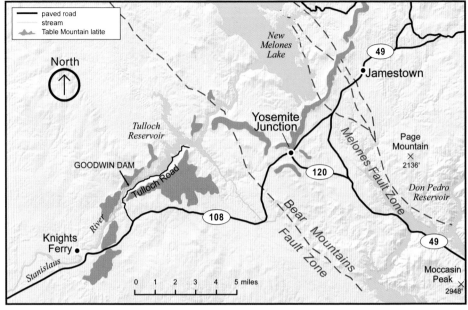

The extent of the Stanislaus Table Mountain lava flow (brown) in the Sierra Nevada foothills. (Modified from Rhodes, 1987.)

GETTING THERE

The Stanislaus Table Mountain lava flow is some 60 miles long, so there are a number of locations from which to view it. Our favorite location is a viewpoint along Tulloch Road (also known as Tulloch Dam Road). To get to Tulloch Road from Big Oak Flat Entrance Station, follow California 120 west for 44 miles to Yosemite Junction and turn left onto California 108/120. Follow California 108/120 west for 10.7 miles and turn right (north) onto Tulloch Road. To get to Tulloch Road from Modesto, follow California 108 east for 30 miles (it becomes California 108/120 in Oakdale). California 108/120 is divided here, so you will need to turn left onto a short connector, then cross California 108/120 west onto Tulloch Road. Proceed 1.1 miles to where Tulloch Road bends rather sharply to the right and goes more steeply downhill. Go another 0.2 mile and park at an obvious wide spot on the left (north) side of the road. This spot is about 0.75 mile downstream of Goodwin Dam, so if you reach the dam you've gone too far. From the turnout, walk a short distance down to a prominent point overlooking the river. Be aware that it is a steep drop to the river here, so keep an eye on children.

Although we don't discuss them in this vignette, other areas to see good exposures of the Stanislaus Table Mountain lava flow include locations near the junction of California 120 and California 108 at Yosemite Junction, along Rawhide Road northwest of Jamestown, at the Parrots Ferry Bridge over New Melones Lake on Parrots Ferry Road, and on Camp Nine Road near the town of Vallecito.

20

Inverted Landscape

THE STANISLAUS TABLE MOUNTAIN LAVA FLOW

The Stanislaus Table Mountain lava flow is one of several volcanic flows on the western slope of the Sierra Nevada. Confusingly, many are named Table Mountain, so we refer to this particular flow as Stanislaus Table Mountain lava flow. This narrow, winding, flat-topped ridge can be traced for more than 60 miles, from near the crest of the Sierra Nevada near Sonora Pass on California 108 to the western edge of the foothills just southwest of Goodwin Dam. It is a spectacular example of what geologists call *inverted topography*, in which former low points of the landscape are now high points.

The Stanislaus Table Mountain lava flow rises above a relatively smooth landscape that geologists call an *erosion surface*, which is a topographic surface that formed as rocks—in this case, those of the early Sierra Nevada—were deeply eroded to form a low-relief landscape (see vignette 21). The rocks that make up this landscape are mostly 200 to 100 million years old and are part of the belt of metamorphic rocks (the Western Metamorphic Belt) that borders the Sierra Nevada on the west (see vignette 19). After the Sierra Nevada Batholith formed 105 to 85 million years ago, erosion dominated the Yosemite region for tens of millions of years, stripping away miles of rock that covered the granite batholith. The metamorphic rocks of the erosion surface in this area are some that survived this erosional planing.

In most places, the Stanislaus Table Mountain lava flow doesn't rest directly on bedrock of the erosion surface, but instead is separated from it by 100 to 300 feet of sedimentary layers, which were deposited on the erosion surface. The lowest of these layers are thin patches of gold-bearing river gravel—the famous auriferous gravel placer deposits of the California gold rush, which were hydraulically mined in the late nineteenth century by washing the gravels through sluice boxes (narrow, artificial passageways) with jets of water. Geologists think these river gravels are around 50 million years old, and presumably they represent river channels that meandered across the old erosion surface.

Sitting above the river gravels are younger deposits consisting of layered volcanic rocks (mostly rhyolites) and sedimentary rocks derived from erosion of these volcanic rocks. Sparse fossilized plant remains and isotopic dating of the volcanic rock suggest that these deposits are about 10 million years old. Volcanic fragments in the sedimentary rocks are similar in composition to widespread lavas at higher elevations in the northern Sierra Nevada. Such volcanic deposits occur in irregular patches across most of the western slope of the range and evidently once formed a continuous sheet that has mostly been eroded away. This volcanic activity blanketed the old erosion surface and river gravels with hundreds to thousands of feet of volcanic sediments.

This package of rocks was not all derived from the Sierra Nevada, however. Yellowish, fine-grained layers seen at the base of it in places are part of the Valley Springs Formation, a unit named for exposures near Valley Springs, about 20 miles northwest of the Tulloch Road site. Rocks of the Valley Springs Formation were deposited by pyroclastic flows—glowing clouds of pumice and volcanic ash produced by colossal volcanic eruptions. That the deposit occurs in the Sierra Nevada is not surprising, but part of the deposit has been traced east across the Sierras to a source near Ely, Nevada, over 300 miles away. That is surprising. Eruptions of this size have (thankfully) not happened for about 73,000 years anywhere on Earth.

About 10 million years ago, a series of volcanic eruptions near Sonora Pass on California 108 produced voluminous lava that flowed both east and west from the crest of the Sierra Nevada. The westward flow, that of the Stanislaus Table Mountain lava flow, was mostly confined to the channel the ancestral Stanislaus River carved into the volcanic rocks mentioned above. The lava flowed downslope for more than 60 miles to the western edge of the Sierra Nevada foothills, filling the channel with 50

Aerial view of the Stanislaus Table Mountain lava flow near Jamestown, looking southwest. The flat-topped, meandering ridge was once a river channel. The body of water is New Melones Lake.

to 300 feet of lava. That the lava traveled such far distances indicates that its viscosity was relatively low, so it was runnier than most of the lava spewed in the region. In contrast, viscous lava builds steep volcanic cones or domes, such as Mount St. Helens or Mount Lassen.

Although volcanism continued off and on in the Sierra Nevada for another 5 million years or so, the Stanislaus Table Mountain lava flow was one of the last major eruptions in this area. After the lava flow cooled, erosion slowly stripped away the surrounding soft blanket of volcanic and sedimentary rocks, reexposing large areas of the old low-relief erosion surface. In contrast, the Stanislaus Table Mountain lava flow proved much more resistant to erosion, and with time it projected higher and higher above the adjacent landscape. The lava flow protected the underlying volcanic sediments, and in some cases the gold-bearing gravels.

Because what was originally a river channel became a ridge, the Stanislaus Table Mountain lava flow is a classic example of topographic inversion. This phenomenon was recognized by geologists over a century ago. William Brewer, one of the first geologists to explore the Sierra Nevada, summed it up in 1862 when he wrote about the Stanislaus Table Mountain lava flow: "This lava, now the top of a mountain, ran in a stream in the bottom of a valley! Those hills of softer slate have been worn down until they are the valleys, while the harder lava has withstood the elements and forms now the top of the mountain."

The overall setting of the Stanislaus Table Mountain lava flow and surrounding area is best seen from the air, but the Tulloch Road viewpoint is a good substitute. From here you can clearly see the lava flow, underlying sedimentary and metamorphic rocks, and the modern Stanislaus River. From this viewpoint, the Stanislaus Table Mountain lava flow appears to consist of at least two flows, with a combined thickness of about 200 feet. The rock is latite, similar in composition to andesite (a common constituent of large volcanoes such as Mt. Shasta) but richer in potassium. In this sense it is similar to the rock at Little Devils Postpile (vignette 18). If you're interested in examining the latite up close, there are exposures on the opposite (south) side of Tulloch Road from where you parked.

At first glance, the Stanislaus Table Mountain lava flow may appear to be the remnant of a vast sheet of lava that once covered the western foothills before it was mostly eroded away, but there are three pieces of evidence that tell us the lava flow has eroded very little. First, the surface of the Stanislaus Table Mountain lava flow is somewhat rough, composed of small knobs and boulders, but overall the top of the flow is remarkably flat. If it had been eroded a great deal, the surface would be dissected by drainage channels. Second, in many places the uppermost lava contains abundant small cavities, called *vesicles*, which formed when bubbles of gas became trapped in the lava. Vesicular lava is most common at the top of lava flows because gas moves upward through the flow and expands when it gets near the surface. That the top of the Stanislaus Table Mountain lava flow preserves vesicles also tells us that the surface has eroded very little since it was erupted 10 million years ago. Third, several miles upstream, near the historic gold mining town of Columbia, the lava flow sits in a small basin that has not been topographically inverted. The width of the preserved channel in that basin (500 to 600 feet) is similar to the width of the preserved channel at points downstream where the topography *is* inverted. This suggests that, like the top surface, the sides of the flow have eroded very little. The narrow, sinuous geometry of the flow essentially preserves the ancestral Stanislaus River channel.

From the Tulloch Road viewpoint, vertical columns are visible in some places near the top of the flow. These columns are defined by joints that formed as the lava flow cooled and contracted. Most columnar joints, as they are called, form perpendicular to the surface that is cooling—that is, they are parallel to the direction heat was flowing. Nearly horizontal lava flows such as the Stanislaus Table Mountain lava flow will generally have nearly vertical columns because the top and bottom of the flow cooled first.

The rock beneath the Stanislaus Table Mountain lava flow (and your feet) at the Tulloch Road location is an intriguing metamorphosed volcanic breccia (a coarse-grained rock composed of angular rock fragments

Columnar jointing near the top of the Stanislaus Table Mountain lava flow at the Tulloch Road site. The columns are about 30 feet tall.

Stanislaus Table Mountain latite

ancestral Stanislaus River gravels

metamorphic volcanic breccia

The view upriver from the Tulloch Road site. The dark, horizontal layers forming cliffs are lava flows of the Stanislaus Table Mountain lava flow. River gravels form the tree-covered slopes below the cliffs, and under them is light-colored metamorphic volcanic breccia that is part of the low-relief erosion surface that formed prior to eruption of the lava.

held together by a fine-grained matrix) that is roughly 175 million years old. It is part of the Western Metamorphic Belt of the Sierra Nevada and most likely was part of an island arc (similar to the Japanese islands) that was plastered onto the western margin of North America by subduction. The fragments of volcanic rock in the breccia have been flattened in a roughly east-west direction, making them slightly elongated in a north-south direction. The tectonic deformation that flattened them also produced the tombstone rocks and wild folds in many of the metamorphic rocks of the foothills (see vignette 19). This breccia is part of the vast, low-relief erosion surface that developed in the western Sierra Nevada prior to the volcanism that began 10 million years ago. Note that the modern river here has cut into the breccia only a few tens of feet. We discuss why later in this vignette.

At this location, it is difficult to determine what the Stanislaus Table Mountain lava flow rests on, for the bottom of the flow is obscured by rock debris shed from the hillslope. However, if you continue upstream

Just downstream of Goodwin Dam the Stanislaus Table Mountain lava flow rests directly on rounded river gravel and cobbles of the ancestral Stanislaus River. The gravel and cobble deposit is about 15 feet thick.

on Tulloch Road for about 0.5 mile to a point not far downstream of Goodwin Dam, you can look across the valley and see the bottom of the lava flow (binoculars will help). Here the lava rests directly on a sedimentary deposit consisting mostly of rounded gravel and cobbles. The ancestral Stanislaus River deposited these sediments prior to the eruption of the latite. In areas upstream, miners tunneled under the Stanislaus Table Mountain lava flow to search for gold in these underlying gravels, without much success.

At this upstream location, you can see that the modern Stanislaus River hasn't cut a very deep channel. In other words, the position of the ancestral Stanislaus River, as marked by the top of the gravels, is not much higher than the position of the modern Stanislaus River. Why would that be, considering the river has been carving its channel for as long as 10 million years? The answer has to do with the nature of the uplift the Sierra Nevada experienced during that period. The Sierra Nevada is thought to have tilted as a block along a northwest-southeast axis that is buried beneath sediments of the Central Valley. The amount of vertical uplift near the axis was small, whereas far from the axis, at the crest of the range, the amount was greater—perhaps as much as several thousand feet. Because the Tulloch Road site sits near the tilt axis, the amount of uplift here was relatively small, and so was the amount of river incision—about 100 feet, judging from this spot.

There are many factors that determine how quickly and effectively a river or stream can erode its channel. Besides the type of rock or sediment being eroded, a river's discharge and gradient (steepness) are key variables. In general, rivers with steeper gradients and larger discharges erode their channels more quickly than rivers with shallow gradients and small discharges. Because uplift was greatest at the range crest, you might expect that the amount of river down-cutting would be greatest there. However, a river can only erode effectively if it runs down a steep slope *and* has sufficient water discharge. The Stanislaus River is only a small stream near the range crest, and it flows for many miles before collecting enough water from tributary streams to have sufficient power to cut deep canyons. As a result, the Stanislaus River canyon is deepest, nearly 2,000 feet deep upstream of Otter Bar, at a point roughly in the middle of the western slope of the range.

The Stanislaus Table Mountain lava flow has been used to investigate the timing and amount of uplift of the central Sierra Nevada. The surface of the lava flow drops between 60 to 300 feet per mile from where it begins near the crest of the range and ends southwest of Tulloch Reservoir. The gradient is considerably steeper than that of the modern river over the same distance, which drops about 25 feet per mile. Geologists have interpreted this difference as resulting from uplift of the Sierra Nevada. Assuming that the ancestral Stanislaus River, with its sediments preserved beneath the lava flow, once had a gradient similar to the modern Stanislaus River, the now steeper gradient of the ancestral river's sediments must have been caused by the westward tilt of the range. Because the Stanislaus Table Mountain lava flow is 10 million years old, a substantial amount of tilt must have occurred after that time. As noted in the introduction, the timing and magnitude of Sierra Nevada uplift is a topic of debate among geologists, and some of the assumptions underlying the debate have recently been questioned. This makes the Stanislaus Table Mountain lava flow an even more important piece of evidence as geologists refine their understanding of the early Sierra Nevada landscape.

The Stanislaus Table Mountain lava flow also preserves evidence of more recent tectonic activity, something that is often hard to find in the relatively old rocks of the Sierra Nevada. Between Knights Ferry and Columbia, the lava flow crosses two major fault zones, the Bear Mountains and Melones fault zones. At these locations the lava flow has pronounced steps down to the east, offsetting the surface of the lava flow, which must have originally been smooth and continuous. The offset indicates movement along these faults zones in the past 10 million years. Although this time estimate isn't terribly precise, if not for the presence of this remarkable lava flow it is unlikely that geologists would have been able to determine whether there had been any movement at all along the faults in the past several hundred million years. Fault offset in younger rocks and sediments is useful information to geologists; it tells them that the faults were recently active and, more importantly, that they might still be active.

The Stanislaus Table Mountain lava flow and other nearby lava flows mark a boundary in the rocks of the Sierra Nevada. From Yosemite south, most of the exposed rocks in the higher part of the range are granite, but from the northern boundary of the park north, much of the granite was buried by lava flows and deposits made of volcanic sediments, like the gravels and cobbles under the lava flow at the Tulloch Road site. If you cross the range over one of the passes north of Tioga Pass—Sonora, Ebbetts, or Carson—you will drive into this striking volcanic terrain. Although these rocks don't erode into the cliffs and domes of Yosemite, they have a charm all their own.

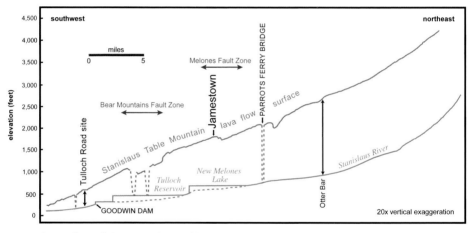

The surface of the Stanislaus Table Mountain lava flow (brown line), as well as the deposits of the ancestral Stanislaus River below it, has a steeper gradient than the modern Stanislaus River (blue line). Assuming the ancestral river and modern river had similar gradients, this difference in gradient suggests that the Sierra Nevada was uplifted, via westward tilting, sometime after the lava flow cooled. The black arrows show the amount the modern Stanislaus River has incised its channel at the Tulloch Road site and at Otter Bar. The steps in the Stanislaus Table Mountain lava flow profile represent offset along faults, and the dashed, vertical lines represent places where the modern river has cut through the lava flow and runs perpendicular to it. (Modified from Rhodes, 1987.)

21

Eocene Erosion

ANCIENT, WEATHERED LANDSCAPES OF THE SIERRA NEVADA

Most visitors are drawn to Yosemite because of the high peaks and spectacular cliffs, and those features have amazing stories to tell. But as is so often true in geology, even seemingly mundane terrain preserves important information about landscape evolution. We present the vertical landscapes of Yosemite in other vignettes; in this one, we'll celebrate landscapes notable for their flatness.

From Rim of the World Overlook you can see both young and very old landscapes. The largest canyon in front of you is relatively young and contains the main stem of the Tuolumne River, which has its headwaters in the Lyell Glacier on Mt. Lyell, the tallest peak in Yosemite National Park. The Tuolumne River has cut the canyon over 2,000 feet below the rim, and it is fairly typical of Sierra Nevada rivers in that it occupies a deeply incised canyon surrounded by relatively flat surfaces rather than sharp ridges. To the right (south) is a major tributary of the Tuolumne River, the South Fork Tuolumne River. After crossing under California 120, the south fork drops more than 1,400 feet through the short but spectacular canyon in front of you, eventually joining the Tuolumne River. The south fork canyon is an excellent example of a hanging valley (see vignette 3). Many hanging valleys have glacial origins, but not all. It's unlikely that the glaciers that once occupied the main Tuolumne canyon ever made it down this far, so the much higher south fork canyon is a hanging valley for another reason, probably having to do with recent uplift of the Sierra Nevada and more intense river cutting in the Tuolumne River canyon associated with that. The steep gradient separating the upper south fork from its confluence with the main Tuolumne River is known as a *knickpoint*, and this knickpoint is probably migrating up the south fork channel. Over time the steep gradient downstream of the California 120 bridge will level out as the gradient upstream of the bridge grows steeper.

The river canyons are steep, but the rims around them are flattish, undulating surfaces. There is a great example of one of these low-relief surfaces in front of you just to the north of the South Fork Tuolumne River. Another, more extensive, low-relief surface makes up the far skyline. Geologists sometimes call these *erosion surfaces,* but the term is somewhat paradoxical here, for these surfaces are actually eroding very

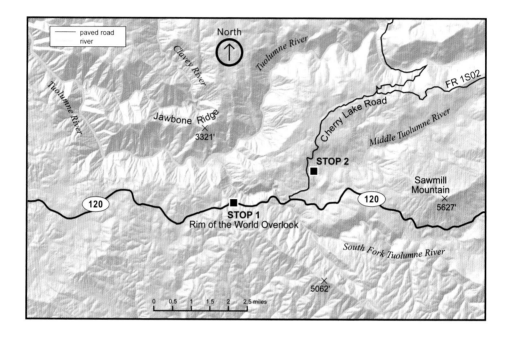

GETTING THERE

Stop 1 is a U.S. Forest Service scenic overlook called Rim of the World. From Modesto, follow California 108 east for 40.8 miles (it becomes California 108/120 in Oakdale). Turn right onto California 120 east at Yosemite Junction and follow it for 17.1 miles. Turn left, staying on California 120 east, and drive approximately 15 miles. To get there from the Big Oak Flat Entrance Station, follow California 120 west (Big Oak Flat Road) approximately 11.8 miles. Look for the large paved parking area on the north side of the road. To get to stop 2 from Rim of the World Overlook, proceed east on California 120 for 1.8 miles and turn left onto Cherry Lake Road (a sign will point out other destinations, such as San Jose Camp). Proceed a little over 1 mile, crossing over the Middle Tuolumne River, until the road starts to climb up a hill with views to the east. Pull off to the side of the road.

slowly. They are called *erosion surfaces* because they represent deep erosion of the rocks of the early Sierra Nevada. Low-relief erosion surfaces are widespread across the Sierra Nevada, extending in broken patches across the length and width of the range. They are more common in the lower parts of the range but are generally more conspicuous in the high country, mainly because at those elevations glaciers dissected the terrain surrounding them, leaving many of them standing as flat plateaus surrounded by deep canyons and high peaks. The plateaus themselves were not covered with ice. A high-altitude example of a low-relief erosion surface is the Dana Plateau, situated east of Tioga Pass near the range crest at an elevation of 11,400 feet. Other well-known examples include the Boreal Plateau, Chagoopa Plateau, and even the summit area of 14,495-foot Mt. Whitney, all located in Sequoia National Park.

How these low-relief surfaces formed remains a topic of debate. The early view held that they formed at low elevations, where meandering rivers beveled them flat, and were subsequently uplifted as the Sierra Nevada rose. In this view, low-relief surfaces at different elevations mark different pulses of uplift. More recent study has revealed that low-relief surfaces can form at high elevations through more subtle processes such as freeze-thaw cycles, in which the progressive freezing and thawing of water in rock slowly breaks it apart and ratchets it downhill. That such subtle processes can, over long time periods, flatten a surface casts doubt on the uplift histories geologists have inferred from the low-relief surfaces. In regard to the Sierra Nevada, a consensus view seems to be emerging that these surfaces are relicts of the ancestral Sierra Nevada, probably

The view to the northeast from the Rim of the World Overlook, showing the main stem Tuolumne and South Fork Tuolumne river canyons incised into low-relief erosion surfaces.

Dana Plateau is an ancient low-relief surface above Tioga Pass that sits at an elevation of 11,400 feet. —Cecil Patrick photo

dating back tens of millions of years to the time when the history of the range was mainly one of erosion, but they need not necessarily have formed at low elevations. Research into this intriguing issue is ongoing.

The erosional processes that affect the plateaus are entirely different from those that erode the adjacent canyons. The canyons have been scoured by rivers and glaciers, forces generally known for eroding landscapes quickly, whereas the plateaus have been eroded by freeze-thaw processes, wind abrasion, and other forces generally known for eroding landscapes slowly. In fact, the low-relief surfaces are very stable parts of the landscape, with astonishingly low rates of erosion; research across the Sierra Nevada suggests that these surfaces are being lowered only a few feet per million years. The canyons, however, are eroding at much faster rates.

A short digression is in order here to clarify how weathering and erosion are different. Weathering is the in-place physical (breaking up) and chemical (altering the chemical makeup) degradation of rock, usually accomplished by water. In granitic terrain, weathering breaks granite down to particles consisting largely of clay, quartz, and some feldspar. Erosion involves the transport of rock debris, by physical forces (rivers, glaciers, wind) or chemical means (dissolving minerals and carrying the chemical constituents away in solution). In most cases, weathering breaks a rock down into smaller particles, and then erosion carries those particles away. In granitic terrain, the weathering and decomposition of granite produces grus, a coarse, sandy material made up of the component parts of granite (quartz, feldspars, and so on).

If an equilibrium between weathering and erosion is established (the rate of weathering equals the rate of erosion), the resulting landscape has a relatively thin mantle of weathered material that doesn't change in thickness over time. However, weathering and erosion aren't always in equilibrium. For example, a bare granite slab must be eroding faster than it is weathering; otherwise, there would be grus on the surface of the slab. Such landscapes are referred to as *weathering-limited* because the rate of weathering is the limiting factor in the overall evolution of the land surface. Many of Yosemite's landscapes are weathering-limited; for example, its bare granite domes are cleared of weathered material as soon as that material is produced. On the other end of the spectrum are landscapes referred to as *erosion-limited*, where erosion rates are slower than weathering rates, so the degraded rock particles aren't carried away as quickly as they are produced. These landscapes are thick with weathering products, such as soil and grus.

Stop 2 is an erosion-limited landscape that has developed on an ancient erosion surface. An obvious feature of the weathered landscape here is the red color of the soil. This red soil, called *laterite*, develops when bedrock undergoes intensive and prolonged weathering and erosion rates are very low. It forms in hot, wet, tropical environments in which chemical weathering is so intense that almost everything, even quartz, dissolves, leaving behind an insoluble mass of iron oxides (hence the rusty red color) and aluminum hydroxides (compounds of aluminum, oxygen, and hydrogen). Obviously, this is no longer a tropical area, and the laterite soil could not have formed in today's climate. This soil records an older tropical climate and probably formed during Eocene time, between 55 and 34 million years ago. The Eocene was characterized by a stable climate that was much warmer, and probably wetter, than today's. Temperate forests, which Yosemite has today, reached the poles, and tropical forests extended as far north as latitude 45 degrees north, the latitude of Maine (Yosemite is presently between latitudes 37 and 38 degrees north). The laterite soil probably developed over tens of millions of years of weathering. Clearly, something as seemingly mundane as soil can preserve important information about past climates.

The fact that the laterite soils are still present after tens of millions of years testifies to the very slow erosion rates experienced by the low-relief surfaces in and around Yosemite. At present, red laterite soils are found exclusively in those parts of the Sierra that were not glaciated, which generally means they are found at lower elevations. These soils probably once existed at higher elevations, but glaciers easily scraped away the soil, along with all other weathering products, such as grus, down to competent bedrock.

Scattered light-colored granite boulders surrounded by red laterite east of Cherry Lake Road. Unlike the powerful main stem of the Tuolumne River, the weaker tributary here hasn't had sufficient time to erode as much of its drainage basin. Thus, the broad valley walls here display deeply weathered surfaces that are relicts of an earlier time.

You will notice that there are a lot of boulders scattered here and there on top of the laterite. Although the contrast between the light-colored boulders and the red soils would suggest that they are composed of different materials, they are both derived from the same granite. The soil is the deeply weathered end-product of prolonged weathering, whereas the boulders are relatively unweathered. How did these boulders come to reside on this surface? It seems unlikely that glaciers deposited them, because glaciers are not thought to have covered this area, and because the boulders are composed of the same rock type as the weathered rock they rest on.

The answer can be found in several roadcuts along this section of Cherry Lake Road. These roadcuts reveal a number of the light-colored granite boulders, called *core-stones*, surrounded by reddish soil. Even though they are typically rounded, suggesting transport by water or ice, which tends to abrade and wear away sharp corners, these boulders were not transported here. Rather, they are weathering in place and owe their rounded shape to the weathering process.

A block of tightly packed sugar cubes is a helpful analogy for understanding how core-stones develop. Within the block, each individual

Light-colored granite core-stones embedded within red laterite soil; both were derived from the same granite.

sugar cube is separated from the others by cracks. Imagine if you sprayed a fine mist of water onto the block. The dissolving power of the water would be focused along the cracks, and with time, the corners of each cube would dissolve and become rounded.

In the case of the roadcuts, the sugar cubes are granite and the cracks are joints (see vignette 10). The joints allow water to penetrate rock masses, and water greatly enhances weathering. Gradually, the rock around the joints weathers, leaving a core of solid rock. As with the sugar cubes, blocks of rock that start off angular are slowly rounded. Rounding decreases a rock's surface area, slowing down the weathering process. Of course, most bedrock joints aren't oriented exactly parallel and perpendicular to each other, as the cracks are in the sugar block analogy, so core-stones aren't uniformly distributed throughout the soil, nor are they always similar in size.

While a core-stone is embedded in grus or soil, groundwater retained within this porous material causes the core-stone to weather rapidly. But once the grus or soil is removed by erosion, the rate of weathering drops markedly because the core-stone is no longer surrounded by wet material.

time/weathering ➤

An idealized schematic of core-stone development. A rock mass begins weathering along joint surfaces, where the edges of joint-bounded blocks are rounded. With time, rounded core-stones (gray) are left in a matrix of weathered, oxidized granite called grus. *With a lot of time and a tropical climate, grus will develop into a red soil known as* laterite.

Even though this boulder near Tamarack Flat Campground closely resembles a glacial erratic, perched as it is on a clean granite surface, this area has not been glaciated. The boulder is an isolated core-stone that was exposed as the grus that once surrounded it was eroded. Whereas the surface must have been erosion-limited in the past, now it is weathering-limited.

As a result, core-stones become concentrated on an eroding surface and commonly roll into drainages and gather there. Called *lag boulders*, these dot the surface along Cherry Lake Road and are concentrated in the small gullies that the road crosses.

Lag boulders can confound geologists who are trying to unravel the history of a landscape. Isolated boulders resting on bedrock surfaces in mountainous environments are commonly interpreted as glacial erratics (vignette 14). There are places in Yosemite, such as the area around Tamarack Flat Campground, where isolated boulders that look just like erratics occur in areas that have never been glaciated. These boulders are core-stones. One way geologists establish whether a boulder is a core-stone or an erratic is to compare the rock types of the boulders to the underlying bedrock. In an ideal world, a core-stone would be composed of the same rock as the underlying bedrock, whereas an erratic would not. However, this is not always the case, as erratics sometimes are composed of the same rock as the underlying bedrock and lag boulders may have rolled onto a different rock type than their source. Thus, isolated boulders in weathered landscapes must be interpreted cautiously.

People often view mountain ranges as either young (for example, the Southern Alps of New Zealand) or old (for example, the Appalachians). But increasingly geologists are coming to understand that a mountain range can contain both young and old landscapes that are effectively separate from each other in terms of their erosion rates and processes. The contrast of the young glaciated landscapes of the Tuolumne Meadows area with the old, deeply weathered erosion surfaces we visited in this vignette indicates that the Sierra Nevada is such a range.

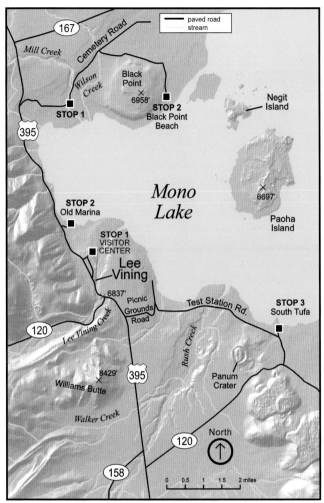

Mono Lake, with the stops for this vignette on the south side of the lake, and those for Black Point (vignette 23) to the north.

GETTING THERE

Mono Lake lies north and east of the town of Lee Vining, which is east of Yosemite National Park. We explore the lake at three stops in this vignette, beginning at the U.S. Forest Service's Mono Basin Scenic Area Visitor Center (stop 1). To reach stop 1 from Tioga Pass Entrance Station, follow California 120 east (Tioga Road) 12 miles. Turn left and follow US 395 north for a little over 1 mile, passing through Lee Vining, and turn right at the visitor center's marked turnoff. The parking lot is about 0.25 mile down this road, toward the lake. The center has interpretive displays, a bookstore, restrooms, and a panoramic view of the lake. (The Mono Lake Committee Information Center and Bookstore, located in the center of town on the west side of US 395, is also well worth a visit. It has an extensive bookstore, an art exhibit, and information about the lake.) To reach stop 2, the Old Marina area, backtrack to US 395 and drive north 1.1 miles. Turn right at the marked picnic area. To reach stop 3, the South Tufa area, from stop 2 follow US 395 south for 7 miles. Turn left onto California 120 east and drive 4.8 miles to the South Tufa turnoff. Turn left (north) on Test Station Road, bearing left immediately after the turn, and drive 1 mile to the parking area. This is a fee area, but it's worth the price.

22

An Ancient, Ice-Bound Sea

MONO LAKE AND ANCESTRAL LAKE RUSSELL

The Mono Lake region is well-known to nature lovers for its spectacular scenery, to bird lovers for its incredibly abundant bird life and importance as a migratory stopover, and to geology lovers for its young and active volcanoes and faults. Most of these topics have been covered in several excellent field guides to the Mono Basin. In this vignette we discuss a subtler, yet easily read aspect: the record of past climates shown by the lake and changes in its water level.

The deserts of the western United States are dotted with hundreds of salty, dusty lakebeds that attest to a much wetter and cooler climate in the past. They were fullest about 20,000 years ago, at or just after the peak of the last glacial period, locally known as the Tioga glaciation, but many were large freshwater lakes as recently as 10,000 years ago. Lakes that form in arid or semiarid areas when the climate gets cooler, or wetter, or both, and then dry up or nearly dry up when the climate gets drier, are known as *pluvial lakes*.

Some of these lakes that formed during glacial times are not quite dry, including Great Salt Lake in Utah, Pyramid and Walker lakes in Nevada, and Mono Lake. Owens Lake was part of this list until 1913, when diversion of water from the Owens River into the Los Angeles Aqueduct caused the lake to dry up. All of these lakes occupy a province known as the Great Basin, which stretches across much of the western United States, from the eastern Sierra Nevada to the Wasatch Mountains in Utah. This region is called the Great Basin because none of the rivers entering it flow to the ocean; instead they pool in the low points and lose water to evaporation. Evaporation concentrates salts and other minerals in these water bodies, and as a result, Mono Lake is roughly twice as salty as seawater. Because their levels depend critically on rainfall and temperature, two climate parameters of vital importance to modern society, such closed-basin lakes are sensitive environmental indicators. Most have gently sloping bottoms, so a change in lake level of a foot might shift the shoreline laterally 50 feet, making subtle changes in lake level easy to spot. For this reason, Mono Lake and other similar lakes have received a great deal of attention from paleoclimatologists, folks who study past climates.

Among pluvial lakes, Mono Lake is particularly useful for climatic studies owing to abundant local volcanism that deposited ash layers in and around the lake. When volcanic ash falls on a landscape, it blankets whatever was at the surface at that time. If it falls in lake water, then it is deposited on lake bottom sediments; if it falls on a mountain, then it might be deposited on bare bedrock; and so on. Because geologists can date volcanic ash, giving us the time at which it was deposited, ash layers provide a time-stamped snapshot of the local geography. Thus, ash layers are one way to determine the extent of lake water in the Mono Basin at various times in the past.

Mono Lake is a salty remnant of a much larger freshwater lake, known as Lake Russell, that occupied Mono Basin during glacial times. Israel Russell of the U.S. Geological Survey worked out the general geologic history of this region in the 1880s. He recognized the ancient, higher shorelines of the lake and the abundant evidence that glaciers flowed into the lake from several canyons west of Mono Lake, including Lee Vining, Lundy, and Bloody canyons. Russell called this larger body of water Lake Mono, but it has since been renamed in his honor.

Russell described the lake and its record of changing climate more poetically than a modern scientist would:

> On looking down on Lake Mono from any commanding point one may easily restore in fancy its leading scenic features at the time of its greatest expansion, as they appeared to the ancient hunters who probably visited its shores. The waters were then fresh and rose several hundred feet higher on the precipitous sides of the mountains along its western border than at present. The peaks of the great Sierra, perhaps then known by names now long forgotten, were white with snow throughout the year and gave birth to ice rivers of great magnitude, some of which reached the shore of the lake. The magnificence of the scene when the Mono Craters were in eruption is beyond description. The ancient sea must have been ice-bound at times for many consecutive years and perhaps for centuries. Again, a change of climate would unfetter its waters and call back the sea-birds to haunt its shore. At all times its scenery was stern and wild and resembled in many ways the grander features of the fiords of Norway at the present day.

At various times Lake Russell overflowed to the northeast, into the Walker River drainage, and to the southeast, into the Owens River drainage via Adobe Valley. Most lakes stick to one drainage outlet, but Lake Russell's outlet shifted owing to faulting and volcanism that rearranged the topography. When the lake spilled into the Owens River drainage, it fed what was, at peak flows, perhaps the most impressive series of lakes that has ever occurred in the western United States. After spilling from Mono Lake, the water ran down the length of Owens Valley and pooled in Owens Lake. From there, the water spilled southward into China Lake and then eastward into Searles Lake. From Searles Lake water spilled eastward into the Panamint Valley, forming the 1,000-foot-deep Panamint Lake. When conditions were right, water spilled eastward over

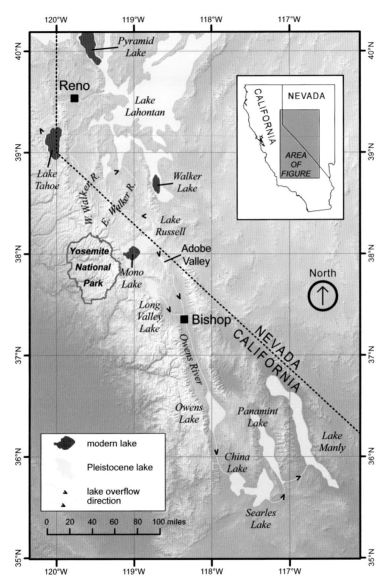

Pluvial lakes that formed at and near Mono Basin during the last glacial period. Some of Nevada's pluvial lakes aren't shown. Lake Russell generally overflowed through Adobe Valley and joined the Owens River near Bishop; during times of greatest runoff water flowed from Bishop south through Owens, China, Searles, and Panamint lakes to Lake Manly in Death Valley. Sometimes Lake Russell overflowed to the northeast, joining the East Walker River to flow into huge Lake Lahontan. (Modified from Reheis and others, 2002.)

Wingate Pass into Death Valley, becoming part of a 100-mile-long lake called Lake Manly. This impressive chain of lakes stretched some 250 miles, from headwaters near Mt. Dana at 13,000 feet above sea level to 282 feet below sea level in Death Valley. As the climate became drier, the lakes were cut off from their freshwater supply and evaporation reduced their levels to below the spillpoints, at which point their shorelines were determined by rates of inflow and evaporation. Evidence for this lake chain remains in the form of ancient shorelines and spillway channels, along with fossils that record connections between various lakes.

Stop 1, the Mono Basin Scenic Area Visitor Center, opened in 1992 in response to growing recreational interest in the Mono Basin. The walkway into the building takes you past a boulder display showcasing some of the common local rock types: granite, rhyolite that weathers to orange, banded pumice, and a great big piece of black obsidian. Amazingly, all of these seemingly different rocks formed from magma of the same composition.

NASA photo of Mono Lake taken by the Landsat 7 satellite on December 16, 1999. Dozens of shorelines left behind as Lake Russell receded from its most recent high level 13,000 years ago are clearly visible, especially northeast (upper right) *of the lake. The transition from smooth to rugged topography on all sides generally marks this high level. Two plumes of gray smoke are visible.* —Courtesy of NASA's Earth Observatory

The granite crystallized at least 1 mile deep in the Earth, whereas the others were erupted at the surface and whipped up to various degrees by dissolved gases. Geologists commonly use beer for analogies, and in this case we can equate pumice to the frothy head on a mug of beer, obsidian to the beer in the mug (a nice dark stout), and granite to beer that has been put in the deep freeze.

Before entering the building or going around to the east side to view the lake, look back toward Lee Vining. The white "LV" on the hillside sits just above a prominent horizontal bench that almost looks like a graded road. This is the 13,000-year-old shoreline, or strandline, carved by wind-driven waves when the lake sat at that level for a long time. The age of this strandline has been estimated from radiocarbon dates of tufa deposited at that level. Tufa is a white calcium carbonate coating precipitated from the soupy lake water (see stop 3 for a discussion of how it forms). The strandline encircles the basin at an elevation of 7,070 feet and marks the highest level that the lake attained since the Tioga glaciation. This bench is about 115 feet below the Adobe Valley spillway, indicating that the lake hasn't overflowed into Owens Valley since then. If it had, the prominent bench would have been erased as it was over-topped by lake water. The "LV" sits on a triangular mountain face, the top of which roughly marks the maximum lake elevation of 7,480 feet. If you mentally extrapolate this elevation around the lake, you can see just how much water Lake Russell contained during glacial times, when it was at least 1,000 feet deeper.

The area behind the visitor center building hosts a patio with an expansive view of the lake. To the left, the steep mountain front of the Sierra Nevada comes down to the western shore of the lake, and US 395 is squeezed between the mountains and the shore. Call this direction nine o'clock. The white area on the far shore is a county park. The broad, dark, flat-topped mound at ten o'clock is Black Point, a volcano that erupted underwater about 13,000 years ago, when the lake was at or near its recent highest level. Dark, hat-shaped Negit Island, clearly a volcano, is at eleven o'clock, and broad, low, light-colored Paoha Island, a less-obvious volcano, is at twelve o'clock. The southern shore of the lake is blocked by a flat-topped mesa; this is the top of a delta that formed when a glacier occupying the valley of Lee Vining Creek dumped into the much higher Lake Russell. The north end of the White Mountains, 45 miles away, shows up above this bench at about one o'clock. Finally, the Mono Craters chain of rhyolite volcanoes is visible on the skyline at about two o'clock.

The abundant volcanoes of the Mono Craters, Negit Island, and Paoha Island show that volcanism has dominated the recent geologic history of the area. Most of these volcanoes are composed of rhyolite lava, which hardens from a light-colored magma rich in silicon and potassium. Rhyolite is compositionally the same as the granite that makes up most of

Telephoto view of the eastern shoreline from a viewpoint overlooking the delta of Rush Creek south of Lee Vining, near the intersection of California 120 east and US 395. The prominent benches mark shorelines of Lake Russell and are visible from the visitor center, especially in the morning, when backlit. The whitish bench halfway up the slope (arrow) is the 7,070-foot shoreline of Lake Russell.

Yosemite National Park. It is the extrusive version of intrusive granite—meaning the rhyolite was erupted onto the surface, whereas the granite cooled below the surface. Several of these volcanoes erupted about 600 years ago, showering the lake with volcanic debris. The most recent eruption in the area occurred on Paoha Island about 250 years ago.

The old Mono Lake marina, stop 2, now serves as a picnic area with a few trees, a few tables, restrooms, and easy access to the lake. With a little detective work, you can discover another interesting facet of the lake's history here. Thousands of boulders lie between the lake's shore and the nearby mountain front. Most are covered with a thin layer of tufa, making them bright white, but where the tufa has been knocked off you can see that most are granitic and metamorphic rocks. The tufa was deposited on the boulders when lake level was high enough to cover them. These boulders match the rocks that make up the steep mountain front, so their presence is easy to explain: they most likely rolled down here from the mountains. There are also several large blocks that eroded from old tufa towers north of the parking area.

Aerial photo, looking north, of Paoha Island (foreground) *and Negit Island, the small dark island behind Paoha. Negit is a volcano composed of abnormally dark dacite lava, a fine-grained igneous rock that has the same composition as granodiorite but differs in appearance (and other qualities) because it was extruded onto the surface rather than having cooled underground. Paoha is mostly composed of lake sediments that were pushed up above the lake surface as magma intruded shallow levels of the crust.*

Approaching the lake from this boulder field, boulders become less abundant, as they should the farther you get from the mountain front; however, as you continue to approach the shore, boulders again become abundant. These boulders are different. They are almost entirely composed of gray rhyolite pumice—a spongy, lightweight rock erupted explosively during a rhyolite eruption—with a substantial tufa coating. They are more difficult to explain than those closer to the highway, as there is no rhyolite uphill from the shore. Another curious feature of these blocks is that they are concentrated along the shoreline at elevations of 6,391 feet and below, and they are especially abundant along the northern shore of the lake, such as at stop 2 of vignette 23.

The best explanation for this mystery is that the boulders formed during an underwater eruption, floated to the lake's surface, and became grounded on the shore when the lake level was 6,391 feet. Like wood, especially frothy pumice floats because it is less dense than water. A radiocarbon date of a pinecone sandwiched between similar pumice blocks in Lee Vining Creek indicated it was about 1,500 years old. Using that date and other clues, geologists estimate the eruption occurred between 1,550 and 1,750 years ago. This is another way in which past lake levels can be placed in time, and it is worth pondering the obvious conclusion that about 1,600 years ago the lake was as low as it is today, without any help from people. When the lake reaches its mandated level of 6,392 feet (see the discussion at stop 3), these blocks will be right at and below the waterline.

One of the main reasons people visit Mono Lake is to see the bizarre tufa towers that emerge from the water at several places along the shore. These towers formed underwater and are now exposed due to lowering of the lake, both natural and artificial. At stop 3, the South Tufa area, there is a fine nature trail that takes you through the towers. Please remember that the towers are quite fragile, irreplaceable, and protected. Stay on the trail and do not touch them.

The tufa towers are composed of calcium carbonate, the same material shells and limestone are made of. They formed when springs discharged freshwater rich in calcium ions into the bottom of the lake. Because the lake is so saline, its water contains a high concentration of carbonate ions. Separately, calcium ions and carbonate ions are soluble in water, but if you put them together they precipitate out of the water as calcium carbonate. This is the same reaction that produces a whitish scum on glassware. The towers are essentially upside-down stalactites, precipitated from the bottom up in water instead of from the top down in air. Israel Russell figured this out by observing submarine tufa towers actively forming near Black Point; they have freshwater discharging from them.

Tufa is not unique to Mono Lake; it forms in many saline lakes and is common in the geologic record of lakes that have long since dried up. Pyramid Lake in western Nevada is ringed by spectacular tufa deposits that are much more developed than those of Mono Lake, and the Trona Pinnacles west of Death Valley are enormous tufa mounds that were deposited under Searles Lake. The particular charm of the Mono Lake tufa towers lies in their slender form, the surrounding stark volcanic scenery, and the relative ease of access.

Since calcium carbonate only precipitates in water, the towers presently out of the water demonstrate that the lake level has fallen in recent times, but how far? Fortunately, there is a long historic record of lake level. Lake level was between 6,405 and 6,425 feet above sea level for a century after record keeping began in 1850. It peaked in 1919 at about 6,427 feet. This peak corresponded, not surprisingly, to a period of abnormally high rainfall. The 6,427-foot level roughly corresponds to prominent strandlines around the lake today and is about 10 feet above the South Tufa parking lot.

Los Angeles began diverting water from streams that feed Mono Lake in 1941 after completion of a system of aqueducts and tunnels that carried water to the Owens Valley and into the Los Angeles Aqueduct.

A solitary tufa tower, about 15 feet tall. Hundreds of birds dot the water, and Paoha Island is in the background.

These tufa islands at South Tufa stand about 12 feet high.

Although lake level initially remained steady, the 1950s marked the start of a precipitous decline that reached its nadir in 1981, at 6,372 feet. At that point the lake's volume had been cut in half and its salinity doubled. Negit Island was no longer an island, allowing mainland coyotes to raid gull nesting sites.

Spurred by concern about these changes and visions of dry, dusty Owens Lake, the Mono Lake Committee was formed. In 1979 the committee and the National Audubon Society sued the Los Angeles Department of Water and Power to stop the diversions. This suit and subsequent legal actions resulted in a 1994 decision that mandates Los Angeles to restore the lake's level to a target of 6,392 feet. This is about 25 feet below the prediversion level but 20 feet above the 1981 low.

If you walk down the boardwalk toward the lake's shore, you can get a feel for what these levels mean. The parking lot would have been under about 10 feet of water in 1920. The base of the kiosk at the start of the paved path is at 6,410 feet, the lake's level in 1951, after water diversion had started. Walking down the path 185 yards brings you to a sign marking 6,400 feet, the lake level in 1959. A drop of only 10 vertical feet caused this dramatic retreat of the lake's shoreline because the slope of

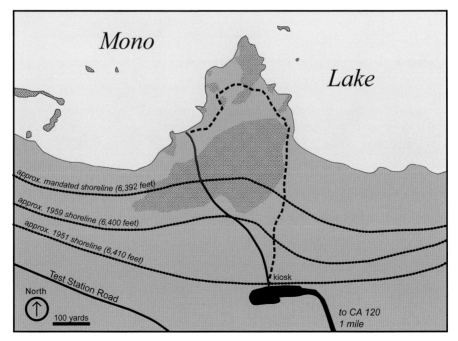

The South Tufa area, showing the paved path (black), boardwalk (gray), and one of many trails (dashed line). As lake level rises, this map will change. The stippling shows areas with significant tufa.

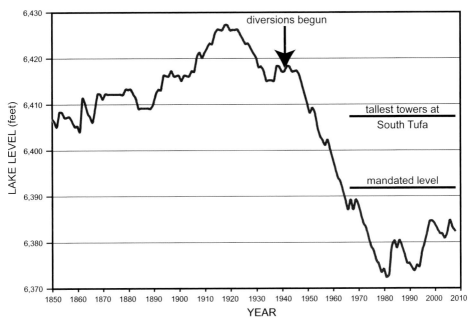

Mono Lake levels (measured on October 1) for the past century and a half, showing variations both natural and unnatural. Lake level was 20 to 25 feet above the 2009 level in 1850, and then rose 20 feet in the next seventy years, all from natural variations in climate. (Data from the Mono Lake Committee.)

this part of the basin is only about 1 degree. Another 130 yards of walking brings you to the end of the paved path and beginning of a boardwalk that leads to the shore. This point marks the mandated lake level of 6,392 feet, and when the lake reaches that level the boardwalk will be underwater.

The tallest tufa towers at South Tufa rise about 25 feet above the water, putting their tops at roughly 6,405 to 6,410 feet. This was at or below the surface of the lake for the century preceding 1950, meaning that most of them were submerged in the first half of the twentieth century. Tufa towers above this level are rare, either because older ones that formed during higher lake levels have crumbled, or because tufa formation is a recent phenomenon. At a few places around the lake, older, higher tufa towers remain. If you go to Black Point on the north shore (vignette 23), look for some around Cemetery Road, about 0.5 mile beyond the county park, at about 6,500 feet above sea level. These lone towers testify to the formation of tufa when the lake was much higher than in historic times.

A stroll around the lakeshore provides fascinating lessons in geology, climatology, limnology, ornithology, entomology, and many other "ologies." The tufa towers come in many shapes and sizes and look different every visit owing to changing light and weather. Look into the lake water near tufa columns and see if you can spot freshwater springs entering the lake—the freshwater looks oily as it mixes with the lake brines and forms new tufa.

Be sure to test the character of the lake water. It is alkaline, with a salinity about 2 times greater than that of ocean water. If you swim in it, you'll float much higher than in ocean water because the high concentration of dissolved salts makes Mono Lake's water significantly denser than ocean water. Be sure to bring some freshwater to rinse off with. Another interesting aspect of this water is that its composition is not unlike a solution of washing soda (sodium bicarbonate), and in fact, many writers have remarked on its ability to clean clothes.

Before you leave the water's edge, look up and try to imagine 1,000 feet of water above your head and how much volume that would be considering the size of the basin. That's how deep and broad Lake Russell was during glacial times. Then consider the evidence presented in vignette 15, which indicates that the lake was far smaller about 1,000 years ago than it is today. These variations reflect natural changes in climate. The Mono Lake area can easily provide a week's worth of study and walking. The Mono Lake Committee Information Center and Bookstore in Lee Vining will provide you with all of the guidebooks and histories you'll need to enjoy this fascinating locale.

23

An Underwater Volcano

MONO LAKE'S BLACK POINT

The region east of Yosemite National Park has experienced regular volcanic activity for most of the last 3 million years and sporadic volcanic activity going back much further than that. This is clearly evident along US 395 south of Mono Lake, where the road passes chains of young volcanoes such as the Mono Craters, but the lake is also bordered on the east and north by extensive fields of relatively young lava. Only on the west is the Mono Lake basin not bordered by more recent volcanism. Although the volcanoes in and south of Mono Lake get most of the attention (see vignette 22 in this book and vignette 30 in *Geology Underfoot in Death Valley and Owens Valley*), the Black Point volcano on the north shore is the most unusual. This volcano and the surrounding landscape are an excellent excuse to visit the northern side of the lake.

GETTING THERE

This vignette explores the Black Point volcano, which dominates the northern shore of Mono Lake, making it a natural add-on to vignette 22 (see the map on page 234). From Tioga Pass Entrance Station, follow California 120 east (Tioga Road) 12 miles. Turn left and follow US 395 north for 5 miles, passing through Lee Vining, and turn right onto Cemetery Road (about 3.7 miles past the turnoff to the Mono Basin Scenic Area Visitor Center). On the north side of the lake you will pass Mono Lake County Park, which provides picnic tables, restrooms, and excellent views of tufa towers, birdlife, and Black Point. You will pass a cemetery 1.3 miles from US 395, at which point the paved road turns to good graded gravel that is suitable for passenger cars. About 0.2 mile later the road crosses Mill Creek and begins a climb. Stop 1 is the roadcut in green sands on your right 0.2 mile after crossing Mill Creek. To reach stop 2, the Black Point access point, continue heading east. After approximately 1.5 miles, just after the road crosses Wilson Creek, bear right at the fork and continue 2.6 miles to the end of the road (this stretch of road is a bit rougher). Though the hike of around 1 mile to the top of the volcano is strenuous, it's worth the effort.

Before we begin our tour of the Black Point volcano, a bit of volcanology is in order. Volcanoes erupt lava flows, liquid masses that flow along the ground, and they also typically blow particles of all sizes into the air (or water). Although the term "volcanic ash" is used routinely in the popular press for much of the material ejected from a volcano, geologists restrict the term *ash* to volcanic particles less than 2 millimeters (about $1/16$ inch) in diameter. Larger particles go by different names, but the term *tephra* encompasses volcanic particles of all sizes. Tephra may pile up around the vent, producing a cone-shaped volcano, or it may travel long distances—even around the world if it is powdery and ejected with tremendous force.

The Black Point volcano is a dark, flat-topped edifice, a little over 1 mile in diameter, that rises about 500 feet above the level of the lake. Flat-topped volcanoes are not unusual (Obsidian Dome and Wilson Butte, which lie just west of US 395 about 18 miles south of Black Point, are examples), but they tend to be made of rhyolite, whereas Black Point erupted basalt. Rhyolite volcanoes can be flat-topped because rhyolite magma, which is extremely viscous, oozes out of the ground, piles up around the vent, and then oozes away a bit to form a pancake shape. Basalt volcanoes, in contrast, generally feed lava flows that flow far from the vent, and any tephra they erupt tends to land close to the vent, building a cone. Basaltic tephra is a spongy-looking, black, rough material of pebble or cobble size. Commercially this tephra is known as *cinder* and is used in gardening and railroad beds. Red Hill north of Inyokern and the volcanoes around Big Pine along US 395 are examples of basaltic cinder cones. So, since Black Point is composed predominantly of basaltic tephra, why is it flat?

Flat-topped volcanoes of basalt can form when there is an eruption under a glacier. Instead of being ejected as tephra and forming a cone, the basaltic lava spends much of its energy melting the ice, becoming a cauldron of lava surrounded by ice and water, and eventually solidifying. When the ice melts, a steep-sided, table-shaped mountain known as a *tuya* remains. Volcanoes of this sort are common in Iceland and British Columbia, where volcanoes have repeatedly erupted under glaciers. There were certainly glaciers reaching down to Mono Basin at times during the past several million years, but the basin itself was never covered in ice, so this mechanism doesn't explain how Black Point formed.

The Pacific Ocean is home to thousands of flat-topped undersea basaltic mountains called *seamounts*, which Black Point resembles. How seamounts formed was a mystery for a long time. Surveying and dredging showed that most seamounts were formerly conical volcanoes that stuck up above the water. Normal erosion doesn't give a conical mountain a flattop haircut; it just lowers the conical profile. However, wave action is efficient at planing off a surface. Geologists deduced that the elevation of conical volcanoes extending above sea level was first lowered due to

The Black Point volcano as seen from Mono Lake County Park. Dark-colored Negit Island is visible in the distance beyond Black Point. Horizontal terraces on the left side of Black Point are former wave-cut shorelines of Lake Russell, a larger version of Mono Lake that existed at the end of the last glaciation.

Aerial view of Black Point; dark-colored Negit Island is in the upper right. Horizontal wave-cut terraces are highlighted by shadows in the lower left, near Wilson Creek.

subsidence, and then, once their tops were near sea level, waves mowed them flat. Continued subsidence caused them to drop below the water's surface. So now we understand how seamounts form, but clearly this mechanism cannot explain Black Point because this region has been on a continent for a few billion years.

Our first clue to the flat-topped origin of the Black Point volcano comes from the remarkable roadcut at stop 1. Your first author nearly crashed his car the first time he drove past this outcrop of wildly folded, dark green layered rocks—it is that remarkable. The green, sand-sized grains are glassy basaltic ash that erupted from the Black Point volcano, and determining how they ended up in such unusual folds takes a bit of detective work.

The layers exposed in this roadcut are part of the Wilson Creek Formation, sedimentary rocks that were deposited at the bottom of Lake Russell, the ancestral version of Mono Lake (vignette 22). The Wilson Creek Formation is composed of a sequence of sandy and muddy sedimentary layers interbedded with nineteen or more light-colored, fine-grained volcanic ash layers derived from the nearby Mono Craters. The youngest beds in the formation are about 12,000 years old, and the oldest more than 50,000 years old. The greenish Black Point bed seen in this roadcut occurs near the top and is about 13,000 years old.

Close inspection shows that the individual green layers are dramatically cross-bedded, meaning individual sand layers truncate other layers, rather than being deposited in parallel sheets. This cross-bedding formed in swift currents that flowed into Lake Russell, which sat at the foot of a tall, glaciated mountain range with active volcanoes nearby. All of this geologic activity left a lot of debris on the landscape that sediment-rich streams carried into the lake.

The amazing pear-shaped layers are folds that formed in the upper sandy layers when they slumped downhill, toward the center of the lake, during earthquakes. How did this happen? The cross-bedded sedimentary layers deposited on the lakebed were saturated and, being parallel to the lakebed, had a slight tilt toward the center of the lake. Earthquake shaking was enough of a jolt to cause some of the layers to slide downhill. The beds were folded and crumpled as they slid across the slippery surface of the beds below. Similar intensely folded layers of sediment occur in sand layers derived from other volcanoes around the lake, indicating that this process was common.

Yet basaltic volcanoes generally erupt cinders that are much larger than the sand grains in this roadcut. This is our first clue to how the flat top of Black Point came about—this sand-sized tephra was erupted underwater. Volcanoes that erupt underwater behave differently than those that erupt in air. The water breaks up the lava, producing abundant sand-sized tephra that is blown about in the huge steam clouds that form as the lava rapidly boils the water.

Intensely contorted beds of basaltic Black Point tephra atop planar beds of the same material exposed at stop 1. The folds developed as the upper layers slumped while underwater—to the right in the photo. The boundary between the planar and folded layers is the slippery surface upon which the sliding took place. The white layer at the base of the tephra (foot level) is composed of a few inches of fine lake sediment (deposited at the bottom of Lake Russell) and is underlain by pinkish ash from the Mono Craters.

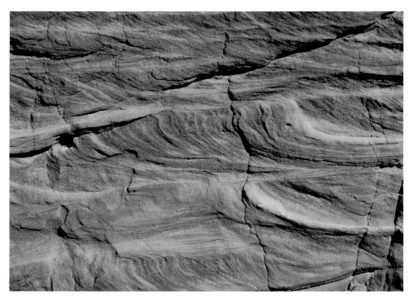

Close-up of cross-bedding in the relatively unfolded layers in the roadcut at stop 1. These particular features, known as climbing ripples, *are produced when swiftly flowing water is laden with sediment. The particles are deposited in waves that build up and migrate in the direction the current is flowing—in this case, from left to right (north to south). When these sediments were deposited, the lake level was much higher and this area was underwater. Ancestral Mill Creek, which emptied into Lake Russell, was carrying glacial sediment from Lundy Canyon as well as volcanic ash from Black Point. The field of view is about 10 inches wide.*

After examining this remarkable roadcut, continue to stop 2. This site offers an unusual vista of the volcanic islands in Mono Lake and is a starting point for hikes to the top of the Black Point volcano. Abundant blocks of tufa-coated pumice litter the shore (described at stop 2 of vignette 22). This area is also much less visited than other areas around the lake, offering a sense of solitude not found at the stops covered in vignette 22.

Radiocarbon dating of organic material just beneath the beds at stop 1 indicates that the Black Point volcano erupted about 13,000 years ago, roughly at the time when Lake Russell was carving a prominent shoreline bench at 7,070 feet (see vignette 22). That means the lake waters at the time were about 110 feet higher than the top of Black Point; thus, the volcano erupted under more than 100 feet of water, confirming the origin of the sand-sized tephra seen at stop 1 and the mechanism that created the volcano's interesting shape.

Although the lower reaches of the hill are covered by sandy soil and vegetation, the upper part of Black Point is composed of light-brown layers of tephra that are roughly horizontal. If Black Point were a normal cinder cone, these layers would dip away from the summit at significant angles because the majority of the tephra blown into the sky would have settled back down close to the vent, parallel to the steep sides of the cone. Because the tephra was erupted underwater, the particles settled back down more slowly and were dispersed by currents, producing gently dipping layers.

Although the view from the top of Black Point is justification enough for the hike, many people make the climb to view deep fissures that cut through the summit. These cracks are up to 80 feet deep, and it's easy to

Basaltic tephra layers near the top of Black Point.

Interior view of one of the largest fissures at Black Point, showing well-developed horizontal layering of basaltic tephra. Ornithologist and budding naturalist for scale.

View to the southwest along one of the fissures at the summit of Black Point, with Mono Lake and the Sierra Nevada crest in the background.

walk into the bottom of some of them. They are coated with tufa, indicating that they either formed underwater or were resubmerged after they formed. Geologists debate the origin of these fissures and have several hypotheses that might explain them: did they form because of rapid cooling when the tephra came in contact with water, because of faults pulling the volcano apart, or because of spreading in advance of magma rising along a dike? The climb to the top is strenuous, especially if the weather is warm, but the view is spectacular. Bring plenty of water and be prepared for a few false summits. Whether you hike to the fissures or just admire the view from the shore, Black Point offers a unique perspective on Mono Lake.

24

Evidence of the Ice Ages

GLACIAL DEPOSITS IN AND AROUND LEE VINING CANYON

Glaciers are largely responsible for shaping the Yosemite landscape. In the introduction, we discussed how glaciers excel at eroding rock, but what happens to the products of this erosion? In this vignette we'll explore some of the deposits glaciers left behind in and around Yosemite. In addition to being interesting features in their own right, these deposits can tell us a lot about the glaciers that left them behind.

The meadows at stop 1, known as Dana Meadows, preserve abundant evidence of the Tioga-age glacier that once occupied Tioga Pass, including piles of till, small recessional moraines (more on these at stop 2), and kettle ponds. The kettle ponds at Tioga Pass are some of the best preserved in all of the Sierra Nevada. Kettle ponds form when a receding glacier leaves behind blocks of ice that become buried by sediment washed out of the melting glacier. When the ice blocks melt, the resulting depressions in the sediment fill with water, forming small ponds. Ponds can also form behind small recessional moraines. The kettle ponds in Dana Meadows appear to have formed in both ways.

As N. King Huber neatly described in his book *Geological Ramblings in Yosemite*, the most intriguing evidence of glaciation at Tioga Pass is the accumulation of "exotic" boulders of Cathedral Peak Granite. This rock type is characterized by large, pink orthoclase crystals (see vignette 1), and because of this, it is the most easily identifiable granitic rock in Yosemite. Boulders of Cathedral Peak Granite are scattered here and there around the kettle ponds in Dana Meadows. If you look carefully, it should only take a few minutes to spot one.

The bedrock around Tioga Pass isn't composed of Cathedral Peak Granite, but rather is composed of older metamorphosed volcanic rocks to the east and the granodiorite of Kuna Crest and the Half Dome Granodiorite to the west. The nearest bedrock outcrops of Cathedral Peak Granite are 4 miles north of Tioga Pass, near Mt. Conness, or an equivalent distance southeast along the Dana Fork of the Tuolumne River. The Cathedral Peak Granite boulders must have been transported to the pass, and this is why Huber considers them "exotic." In fact, they are excellent examples of glacial erratics (see vignette 14), another common deposit of glaciers.

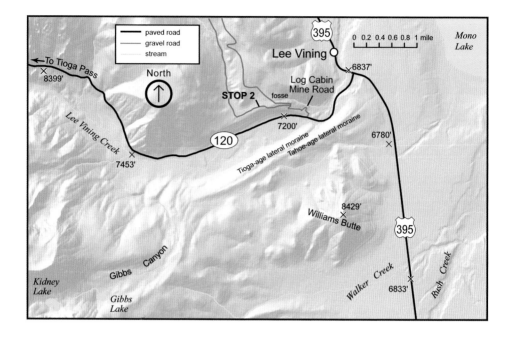

GETTING THERE

Stop 1 is near Tioga Pass Entrance Station. From Big Oak Flat Entrance Station, follow California 120 east (Big Oak Flat Road) 7.8 miles. Turn left, continuing on California 120 east (Tioga Road) approximately 46.6 miles. From the town of Lee Vining, follow US 395 south approximately 0.5 mile. Turn right onto California 120 west (Tioga Road) and drive 12 miles. Park in the small parking area just south and east of the park entrance station and walk south across Dana Meadows to view the area's glacial deposits and ponds. Stop 2 is in Lee Vining Canyon. There are numerous viewpoints along California 120 (Tioga Road) in Lee Vining Canyon from which to view massive glacial moraines, but stop 2 offers an impressive overview. From stop 1, follow California 120 east (Tioga Road) approximately 11 miles and turn north (left) onto Log Cabin Mine Road (graded dirt), which is directly across from a U.S. Forest Service ranger station. Note your mileage. Proceed up the steep road, keeping left at the first junction 0.2 mile from the junction with California 120. This junction is at the top of the Tioga terminal moraine. Turn left at a second junction at 0.6 mile. At 0.7 mile there will be a low bench off to your left, which marks the top of a Tioga-age lateral moraine. Continue to the top of the hill, which is 1 mile from the California 120 junction, and park in a small turnout on the left. From here you have a commanding view to the south of lower Lee Vining Canyon and its impressive moraines. To reach stop 2 from Lee Vining, follow US 395 south for 0.5 mile and turn right onto California 120 west. Drive approximately 1.2 miles and turn north (right) onto Log Cabin Mine Road.

Kettle ponds in Dana Meadows as viewed from the Gaylor Lakes Trail.

How did these Cathedral Peak Granite boulders get here? Tioga Pass is presently the drainage divide between the westward-flowing Tuolumne River and eastward-flowing Lee Vining Creek. However, it appears that about 20,000 years ago, during the Tioga glaciation, the arrangement was somewhat different. Remember that glaciers flow according to the slope of their surface, not the slope of the ground beneath them. This is one of several key ways in which glaciers differ from rivers, and it allows glaciers to flow uphill for short distances (see vignette 13). Ice actually flowed uphill to spill eastward over many of the prominent passes along the Sierra Nevada crest. This happened because there was generally a greater accumulation of ice west of the range crest than to the east, so the ice divide (the point separating the directions ice flowed) was positioned west of the present drainage divide (the point separating the directions the rivers flow).

The Cathedral Peak Granite boulders at Tioga Pass are most abundant at the pass itself and decrease in abundance south and west through Dana Meadows, suggesting that the southwest end of Dana Meadows is farther from the source of the boulders than the northeast end. Thus, the field evidence suggests that these boulders were derived from somewhere north of Tioga Pass. Reconstruction of the Lee Vining glacier, which originated in the Mt. Conness–Saddlebag Lake area, indicates that it was more than 1,000 feet thick where Lee Vining Creek turns east below Tioga Lake and flows into Lee Vining Canyon. As this location is only about 500 feet lower than Tioga Pass, a 1,000-foot-thick glacier could easily have sent an arm flowing south, up and over that pass, as well as a larger arm flowing east down Lee Vining Canyon toward Mono Lake. As long as the surface of the Lee Vining glacier sloped down to Dana Meadows from the Mt. Conness–Saddlebag Lake area, it would have had no problem flowing over the modest rise that is Tioga Pass. In fact, the pass itself is composed of heaps of till, which were likely deposited there during deglaciation (the kettle ponds have formed in this till), so the elevation of the pass may have been somewhat lower during glacial times. Your second author has traced the maximum elevation of Cathedral Peak Granite boulders on the hillsides above Tioga Pass. The highest boulder sits about 200 feet above the present elevation of the pass, indicating that the ice must have been at least this thick as it flowed south over the pass.

It is fortunate for geologists that a rock as distinctive as the Cathedral Peak Granite composes the core of Yosemite, for it provides an excellent tracer of glacial coverage and flow paths. Boulders of Cathedral Peak Granite can be found far from their bedrock outcrops, including

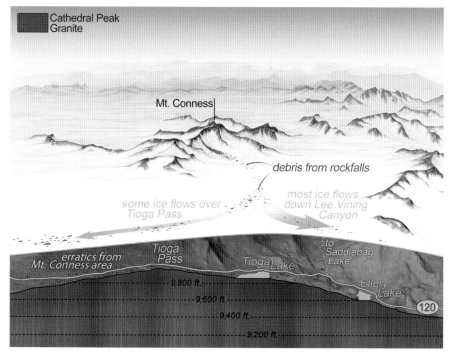

Schematic illustration of the glacier that surmounted Tioga Pass between Lee Vining Creek and Tuolumne Meadows. This flow of ice carried boulders of Cathedral Peak Granite from the vicinity of Mt. Conness, northwest of Tioga Pass, to the meadows south of the pass.
—Illustration by Eric Knight

View to the northeast from the slopes east of Tioga Pass. The large, rounded boulder in the foreground is composed of Cathedral Peak Granite. A glacier carried it to this spot from upper Lee Vining Canyon, visible in the far distance. This boulder sits about 200 feet above the present elevation of Tioga Pass and represents the minimum thickness of the ice that flowed south over the pass.

in Poopenaut Valley in the lower Tuolumne River canyon, on the Diving Board on the west shoulder of Half Dome, in the El Capitan and Bridalveil Meadow moraines in Yosemite Valley (see vignette 9), and at Turtleback Dome on the southwestern rim of Yosemite Valley. Without these boulders, many subtle deposits in Yosemite could not be reliably identified as having a glacial origin. We therefore salute the ever-helpful Cathedral Peak Granite.

Most of the debris glaciers leave behind is in moraines—ridges or lobes of jumbled boulders, gravel, and sand that can be tens to hundreds of feet high (see the introduction for more details). There are many fascinating moraines in Yosemite, but Lee Vining Canyon, between Tioga Pass and Mono Lake, has some of the Sierra Nevada's most spectacular.

Glaciers deposit moraines along their margins where melting occurs. When the amount of ice flowing from the highlands equals the rate of melting occurring at a glacier's margins, the glacier's edges don't shift position, and this allows debris carried by the ice to pile up along the edges. The moraine that forms at a glacier's terminus, or end, is called a *terminal moraine*. These represent the farthest advance of a glacier.

Debris that is piled along the side of a glacier is called a *lateral moraine* (when looking down-valley, the moraine on the left side of a valley is called the *left-lateral moraine*, and the one on the right is the *right-lateral moraine*). As the climate warms and accelerated melting causes the terminus of a glacier to retreat upslope, periodic pauses in the retreat are marked by recessional moraines.

From Tioga Pass, proceed to stop 2, which offers a sweeping panoramic view of lower Lee Vining Canyon from its highest left-lateral moraine. The lateral moraines begin in earnest at the edge of the mountain front and extend out beyond it several miles. In lower Lee Vining Canyon, these moraines tower up to 700 feet above the valley floor. The glaciers that formed these moraines originated from large ice fields that formed at different times over the range crest. Most of the ice that formed in upper Lee Vining Canyon flowed south and east down the canyon, but some went south over Tioga Pass, as mentioned at stop 1, and some even flowed north into upper Lundy Canyon.

The moraines in lower Lee Vining Canyon preserve evidence of at least two glacial advances, the younger Tioga advance (which peaked about 20,000 years ago) and the older Tahoe advance (which peaked between 140,000 and 80,000 years ago). A glacial advance is a period during which a glacier grows and its terminus is moving forward. The Tahoe

Aerial view of the moraines in lower Lee Vining Canyon, looking east, showing lateral, terminal, and recessional moraines of at least two different glacial advances.

lateral moraines form the outer, uppermost ridges, and the Tioga lateral moraines are inset within them. Because the Tioga moraines are inset within the Tahoe moraines and are smaller, we know they were deposited more recently than the Tahoe moraines, and by a smaller glacier. They can't be older because the glacier that deposited the larger, outer Tahoe moraines would have wiped them out. In addition to their relative position, Tioga moraines can be distinguished from Tahoe moraines by their generally sharper crests and unweathered boulders. The formerly sharp crests of the Tahoe moraines have been rounded with time, and many of the boulders they contain have obviously been weathering at the surface for quite some time. Stop 2 is on the top of the highest Tahoe left-lateral moraine, and though there are still plenty of boulders on the surface, you should be able to detect differences between the degree of weathering on this moraine and the inset Tioga moraines below.

There is a lush little valley on the north side of the road running parallel to the Tahoe moraine. Known as *fosses*, small valleys like these form as a tributary stream entering a glaciated (or formerly glaciated) valley is diverted by a lateral moraine. (The troughs between adjacent lateral moraines from two glacial advances are also called fosses.) The stream parallels the lateral moraine for some distance before cutting through the moraine or, in the case of Lee Vining Canyon, joining the main valley below the terminal moraine. Even during the height of the Tahoe glaciation, the fosse was probably an oasis of sorts, harboring vegetation and creating wildlife habitat along the edge of the otherwise inhospitable glacier. Although the glacier is now gone, the fosse here continues to be a pleasant place.

The probable Tioga terminal moraine can be seen in a roadcut just east of the turnoff to Log Cabin Mine Road (1.1 miles west of the junction of California 120 and US 395). This is very likely a terminal moraine, not a recessional moraine, because Tioga-age lateral moraines end at this terminus position. There are at least six recessional moraines farther up the flat floor of lower Lee Vining Canyon. Most are visible on the south side of California 120, and nearly every slight rise in the road through this stretch occurs where the road traverses a recessional moraine. (A particularly large one is 2.9 miles west of the junction of California 120 and US 395.) The Tioga recessional moraines in Lee Vining Canyon are roughly similar in size and shape to the recessional moraines in Yosemite Valley, such as the El Capitan moraine (vignette 9), and formed in the same way.

During the Tioga glaciation, the Lee Vining glacier apparently didn't quite reach the shoreline of Lake Russell, the larger, ancestral version of Mono Lake (vignette 22). During Tioga time the lake was at an elevation of about 6,700 feet (it subsequently rose to 7,070 feet during deglaciation, still not quite high enough to reach the elevation of the presumed Tioga terminal moraine). It seems likely that the glacier didn't reach the

The California 120 roadcut through the probable Tioga terminal moraine near the eastern end of Lee Vining Canyon, just east of the turnoff to Log Cabin Mine Road (1.1 miles west of the junction of California 120 and US 395). The moraine is composed of an unconsolidated mix of silt, sand, gravel, and boulders.

lake; if it had, a terminal moraine wouldn't have formed because the material deposited at the glacier's terminus would have entered the lake and been washed away by wave action.

However, during the earlier Tahoe glaciation, the Lee Vining glacier extended about 1 mile farther down the canyon, as evidenced by the longer Tahoe lateral moraines, and by the fact that there is no recognizable Tahoe terminal moraine. Lake Russell was also larger during Tahoe time. Thus, rather than forming a large terminal moraine, the Tahoe glacier instead calved into the lake, sending out flotillas of icebergs. Lee Vining Canyon and Lake Russell must have been quite a sight at the height of the Tahoe glaciation. If there had been cruise ships back then, a popular activity would have been watching sediment-laden icebergs calve off the glacier into Lake Russell. Today, the best we can do is examine the Tahoe-age lateral moraines near the junction of California 120 and US 395, which are cut by Lake Russell shorelines, forming large benches composed of glacial sediment that was reworked by waves. These benches scribed into the moraines represent different lake levels. Because these wave-cut benches are some of the few flat areas on the edge of the mountain range, they caught the eye of developers and now support a gas station, deli, and other buildings. Once you realize what these benches are, it isn't too hard (except perhaps on an especially hot day) to imagine Lee Vining Canyon full of ice and debouching into a much larger lake.

One of the mysteries regarding moraines along the eastern escarpment of the Sierra Nevada is why they appear to be so much larger than their counterparts on the western slope. Given that Tioga-age glaciers on the western slope were, on average, seven times longer than those on the eastern slope, it would make sense that the larger western glaciers would

Half
Dome

Mt.
Lyell

TUOLUMNE ICE FIELD

Tioga
Pass

Grant glacier

Lee Vining glacier

Lundy glacier

Mono Lake

Lake Russell as it might have looked during the Tahoe glaciation. Glaciers flowed from the vast Tuolumne ice field down the eastern escarpment of the Sierra Nevada and calved icebergs into Lake Russell. Present-day Mono Lake shown for scale.
—Illustration by Eric Knight

have left behind larger moraines, yet this apparently is not the case. We have spent a lot of time speculating on why this might be, without any epiphanies. What follows are some possible explanations.

First, it may be that the lateral and terminal moraines bounding the western-slope glaciers were in fact larger at the end of the Tioga glaciation but have since been mostly destroyed by erosion. Erosion rates tend to be faster on the western slope of the Sierra Nevada because it receives much more precipitation than the eastern slope. Different topographic settings also likely play a role. A primary difference between the eastern and western glaciers is that the eastern glaciers extended beyond the mountain front and into the adjacent basins and valleys. As such, the lateral and terminal moraines extend out from the mountains and actually make up local high points in the landscape. Furthermore, the eastern canyons typically contain small streams without sufficient power to move much material. In contrast, the western glaciers never reached the mountain front, but instead terminated roughly midway up the slope. At their terminus positions, these western glaciers and their moraines were deeply ensconced in the rugged canyons of the major westward-flowing rivers. River down-cutting (and associated hillslope failures) probably scoured out these moraines, washing the sediment downstream to the Central Valley.

Second, in at least some cases the apparent size difference may be simply that: apparent. For example, the Bridalveil Meadow moraine (see vignette 9), the presumed Tioga terminal moraine in Yosemite Valley, projects only about 35 feet above the Valley floor, less than the Tioga terminal moraine in Lee Vining Canyon. However, the actual depth to

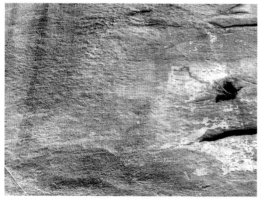

Glaciers had an easier time eroding the densely fractured granodiorite at the crest of the Sierra Nevada near the Dana Plateau (right) *than the massive, unfractured, and polished Cathedral Peak Granite on the western slope of the Sierra Nevada near Tenaya Lake* (left). *This may account for the large size differences between moraines on the eastern and western slopes. Note the bushes for scale; in both photos the view is about 100 feet across.*

bedrock in this area is thought to be quite large, on the order of 650 feet. It is unlikely that the Tioga glacier was scraping along bedrock in Yosemite Valley (also discussed in vignette 9), but it is reasonable to assume that it was moving along a surface that was somewhat below the modern level. If this is correct, the Bridalveil Meadow moraine may actually be much larger than is apparent, with only the top of the moraine poking up above the sediments that have been deposited in the Valley since the glaciers melted.

Finally, the size of the eastern and western moraines may have something to do with the erodibility of bedrock. Most of the glaciers flowing eastward from the crest of the range were eroding canyons cut into granitic and metamorphic rocks. The metamorphic rocks tend to be highly deformed and fractured, having been put through the tectonic "rock crusher" more than once. Granitic rocks near the crest and on the steep eastern slopes also tend to be more jointed and fractured than the granites farther west, perhaps because they lie close to the major faults that fracture and define the eastern edge of the Sierra Nevada, and upon which recent uplift occurred. Glaciers have a relatively easy time eroding fractured rocks because they can pluck out large blocks. Where the rocks are massive, as are the granites of the western slope, glaciers have a harder time plucking blocks and instead mostly abrade and polish the rocks. Thus, the large moraines on the eastern slope may owe their size to the abundance of fractured rocks there.

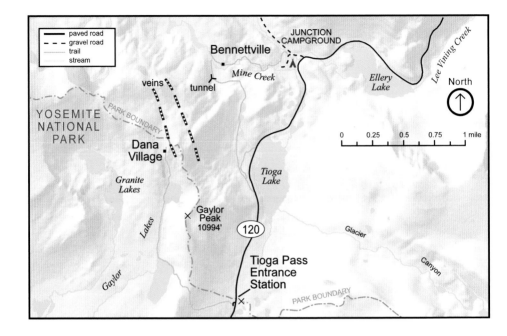

GETTING THERE

Bennettville and Dana Village are located near Tioga Pass. To reach stop 1 (Bennettville) from Tioga Pass Entrance Station, follow California 120 east (Tioga Road) 2.2 miles and turn left (north) onto Saddlebag Lake Road. To reach stop 1 from Lee Vining, follow US 395 south for 0.5 mile and turn right onto California 120 west. Drive approximately 9.8 miles and turn right onto Saddlebag Lake Road. Park 0.1 mile from the turnoff and walk west on the bridge across Lee Vining Creek to Junction Campground. Stay right before entering the campground and follow the signs to the trail to Bennettville, then follow the trail approximately 0.3 mile to the wooden buildings. The mine's tunnel entrance is about 900 feet to the southwest. A convenient loop trail heads northwest from the buildings, crosses the outlet stream of Shell Lake, and then curves back south to visit the tunnel. From the tunnel you can double back or take the old gravel road south to Tioga Lake.

Stop 2 (Dana Village) is reached by a short, strenuous, and rewarding hike from Tioga Pass. From stop 1, backtrack 2.2 miles on California 120 west (Tioga Road) and park in the small parking lot on the west side of the road just south of the entrance station. Hike 0.5 mile west on the Gaylor Lakes Trail to the summit of the ridge, which offers a commanding view of the meadows around Tioga Pass and the Dana Fork of the Tuolumne River. Continue west a few tenths of a mile, dropping steeply to the shore of the largest of the Gaylor Lakes, then follow the trail north about 1 mile to abandoned stone buildings and other mine workings.

25

Dreams of Silver
THE MINES OF BENNETTVILLE AND DANA VILLAGE

As mountain ranges go, the Sierra Nevada is rather barren of minerals. Although the great gold rush of the mid-1800s drew many settlers to California, most of the gold was scavenged from river sediments and there are few mines within the range itself. This is one reason that so much of the Sierra Nevada is still wilderness. Unlike the Rockies, which are richer in mineral deposits and therefore crossed by thousands of old mining roads, the Sierra Nevada has remained largely untouched, except at the margins.

Evidence of old mining activity in the Sierra Nevada is generally found in the metamorphic rocks that surround the granitic plutons, where valuable minerals, known as *ore*, were deposited. While the plutons were cooling, their heat generated convection cells that cycled water through the metamorphic rocks over and over again. This is the sort of hot-water cycling seen in geothermal areas such as Yellowstone National Park, where magmatic heat warms groundwater, which then rises as hot water or steam. This heated water may cycle back down after it cools, or it may escape to the surface in hot springs and geysers. The system is recharged by rainfall. If chemical conditions are right, over long periods of time this water circulation can concentrate precious metals in cracks in the rock. These mineral-filled cracks, called *veins*, may form when a rising fluid with metals in it cools down or when two different fluids meet and chemically react. Regardless of how it happens, veins concentrate metals that had been dispersed over a wide volume of rock into much smaller, minable volumes. Much of human history is concerned with, and many wars have been fought over, such ore deposits, and how and where to find them occupies a significant niche in geology.

The story of Bennettville and Dana Village is told in the highly entertaining history *Bennettville, and the Tioga Mining District*, by Alan Patera. We will only touch on the historical highlights of these fascinating settlements and instead concentrate on the beautiful geology visible at the sites.

The first signs of silver in the Tioga Pass area were recognized in 1860 by a group of prospectors, but they soon left the area for rumors of richer ore at Aurora, Nevada. The vein of silver ore near Bennettville was rediscovered in 1874 by a young shepherd, William Brusky Jr., and has since been named the Sheepherder Lode. Sometime during the winter of 1877–78 the ore was assayed and found to contain significant silver, and many miners staked claims in the Bennettville area, including Brusky, who staked four. A second vein containing silver was discovered west of the Sheepherder Lode, and it became known as the Great Sierra Lode.

Mining occurred at Bennettville and at Dana Village, which lies about 1 mile southwest of Bennettville, on the other side of the ridge that miners tunneled into. Both ore deposits were steep, quartz-rich veins that developed in metamorphic rocks (mostly schists and slates) just east of the main bodies of granite and granodiorite that make up most of the park. The prominent layering that schists exhibit is called *foliation*, and if you think of the foliated schists of this area as the pages of a book that stands vertically, the ore zones are like silver-rich chapters.

Getting ore out of a vein that goes down into the ground is problematic because it requires hauling all of the rock out of the deepening shaft and pumping out the water that seeps in. As the shaft gets deeper, the effort, expenses, and hazards involved increase rapidly. The miners at Bennettville decided to drive a horizontal tunnel into the ridge to avoid these problems. This way they could remove ore and waste rock with cars on tracks, and water could drain out horizontally or be pumped out more easily. They planned on intersecting both veins: the Sheepherder Lode about 1,800 feet in and the Great Sierra Lode, which was also being developed at Dana Village, about 2,800 feet in. Once the veins were struck, the miners would be able to tunnel along them horizontally, up, or down.

The only problem was getting the required machinery up to Bennettville without roads. A lot of machinery was needed: a steam engine, a boiler, a compressor, drills, and other heavy gear—16,000 pounds in all. The miners decided to bring the equipment in from the nearest accessible place, Lundy Canyon, which lies to the north, and they decided to do it in winter. Sliding the freight over snow was deemed easier than carrying it over trackless rocks. In the winter of 1882 they succeeded in moving the equipment from Lundy Canyon up into Lake Canyon, a steep ascent of nearly 1,000 feet, and then up and over another divide into Lee Vining Canyon and on to Bennettville.

This was all well and good, and the boosters of the project thought they might eventually have a town of fifty thousand, replete with a post office and other amenities. A wagon road west to Big Oak Flat (the ancestor of today's Tioga Road) was constructed in 1883. All seemed on track for prosperity, but after nearly 2,000 feet of tunneling no ore had been found. Then the money ran out, and the mine closed in 1884.

At stop 1, the short hike from Junction Campground takes you through the forest to a pair of buildings that mark the town site. Although the town dates from the late nineteenth century, the buildings are in unusually good repair because the Forest Service restored them in 1993. Take

Bennettville in its prime, date unknown. —Courtesy of the National Park Service

care in crossing streams, as high water can persist until midsummer in this area. Late summer and fall are good times to visit.

The rocks at Bennettville are tightly folded, brown metasedimentary (metamorphic sedimentary) rocks. They make an interesting contrast with those of the nearby May Lake area, discussed in vignette 17, which are largely metamorphosed, coarser-grained sandstones, some of which are brilliant white. These rocks were deposited between 500 and 400 million years ago as mud, which was subsequently buried and metamorphosed, forming schist and slate. They are similar to rocks to the east, in Nevada, and demonstrate that this part of the world was deep ocean several hundred million years ago. The veins that were of such great interest to the miners came much later, after the granite and granodiorite plutons of Yosemite invaded the area 105 to 85 million years ago.

Because they were smoothed and polished by glaciers, folds in the Bennettville rocks are spectacularly displayed, especially on the flat areas west of the two wooden buildings. Beautiful chevron (zigzag) folds occur here, as well as boudinage—layered rocks that were pulled into pieces as weaker rocks flowed into the gaps around them. The pieces of ripped-apart rock, which resemble sausages, speak to the tremendous amount of tectonic deformation these rocks endured. Look for pieces of rusty red, quartz-rich rocks that were pulled apart within greenish, finer-grained rock. There are many beautiful rocks to see in this small area, including

Glacially polished folds in metamorphosed mudstone (slate) at Bennettville. Tight V-shaped folds such as these are known as chevron *folds. Knife for scale.*

The tectonic deformation of the rocks around Bennettville was intense. Here, silty, brownish slate layers were pulled apart as the weaker grayish rock flowed around them, forming features geologists call boudins. *Boudin is the French word for "sausage"—in this case, tasty reddish and brownish ones. Penny for scale.*

abundant light-colored glacial erratics of granite and granodiorite that dot the brown surface of the metamorphic rocks.

After exploring the deformed rocks around the wooden buildings, cross the small creek that drains Shell Lake and walk west to the base of the slope. The trail takes you to the abandoned tunnel driven westward into the rock. The hazardous tunnel is fenced off.

After exploring Bennettville, drive to Tioga Pass and hike to the site of Bennettville's sister city, Dana Village (stop 2), which is marked by the remains of stone buildings, iron equipment, and many mine shafts and prospect pits. The hike over the ridge west of the pass is quite scenic and takes you along the contact between granites of the Tuolumne Meadows area and metamorphic rocks to the east. The rocks around Dana Village are somewhat different from those at Bennettville. They are largely gray metamorphosed volcanic rocks between 225 and 165 million years old. Some, including the gray rock that makes up most of the bedrock in the area, were erupted from a caldera near Mt. Dana. These rocks represent some of the earliest volcanism in the Sierra Nevada region. They are strongly foliated, with layering that is mostly planar, rather than wildly folded like some of the rocks at Bennettville. This texture developed as the rocks were heated and squeezed during metamorphism.

The tunnel entrance at Bennettville, with abandoned heavy equipment. These engines would be easy to bring up here today, but in 1882 it was quite a task.

Rocks in this area attracted the attention of prospectors because in places they are soft and crumbly, contain many quartz veins, and occur in shades of yellow, orange, and tan. These characteristics are associated with ore deposits because they are a result of the chemical changes that accompany the formation of ores. Look for small, rusty, cubic holes a few tenths of an inch across in these rocks; these are places where cubes of pyrite have weathered out. Pyrite is another clue that ore might be found in a rock, because it is associated with both gold and silver. Weathering of pyrite, which is a sulfide of iron, produces sulfuric acid, which dissolves and colors the rocks.

Dana Village's desolate stone foundations attest to the hard work and high ambitions of the miners who attempted to establish a high-altitude town within what is now Yosemite National Park. Its remoteness underscores the problems early miners had in developing prospects. Imagine the work involved in digging the tunnels and shafts around you by hand when there were no roads anywhere in the area. Although some silver was recovered, it wasn't enough to cover the costs involved in developing the mines. These empty pits are just one example of the failures that have characterized much of the history of mining in the West.

Gray, metamorphosed volcanic rock, about 222 million years old, at Dana Village. The texture of this rock reveals that it was a pyroclastic deposit, formed when frothing pumice was ejected from a volcano in a huge eruption. The dark gray streaks are lumps of pumice that were flattened when the deposit settled, and then smeared out when the rock was deformed. The lighter gray material was fine ash erupted along with the pumice.

The remains of a stone building and its chimney at Dana Village during an October dusting of snow, looking south. The striking two-tone peak in the background is Gaylor Peak.

View to the northwest along an old prospect pit (the depression in foreground) at Dana Village, looking along the foliation of the metamorphic rocks in this area. The alignment of the foliation indicates that the rocks were deformed by stress that squeezed them perpendicular to the foliation (see the explanation of foliation in vignette 19).

abrasion. The mechanical wearing or grinding away of rock surfaces by the friction and impact of rock particles transported by wind, waves, ice, running water, or gravity.

accumulation. All processes that add snow or ice to a glacier, including snowfall, avalanches, and snow transport by wind.

andesite. A fine-grained, dark, primarily extrusive igneous rock that is richer in silicon and potassium than basalt.

aplite. A fine-grained white rock with a sugary texture that is found as thin dikes in granite and related rocks.

ash. See **volcanic ash**

basalt. A fine-grained, dark, primarily extrusive igneous rock that is relatively rich in calcium, iron, and magnesium and relatively poor in silicon.

batholith. A large (mountain-scale) agglomeration of plutons.

bedding. The layered structure of sedimentary rocks.

bedrock. Relatively solid rock underlying a mantle of loose rock detritus.

biotite. A common rock-forming mineral of the mica group, usually black or brown, shiny, flexible, hexagonal, with cleavage that allows it to flake into thin sheets.

boulder. A rock fragment that is larger than 10 inches in diameter and usually worn and at least partly rounded.

breccia. A sedimentary rock consisting of angular rock fragments held together by a mineral cement or fine-grained matrix.

calcite. A widespread, abundant mineral composed of calcium carbonate. It is the major component of limestone and marble.

caldera. A large circular or oval basin that forms where the ground collapses following a voluminous volcanic eruption.

Cathedral Peak Granite. A large granite pluton that makes up much of eastern and northern Yosemite National Park and is the innermost main unit of the Tuolumne Intrusive Suite. It is characterized by coarse grain size and huge crystals (megacrysts) of orthoclase.

chlorite. A dark green mica-like mineral that is characteristic of low-grade metamorphism.

cinders. Glassy, porous, pea- to baseball-sized fragments of lava ejected explosively from a volcanic vent.

cirque. A wide, steep-walled, flat-floored, semicircular topographic basin created by glacial excavation high in mountainous areas.

clay minerals. A family of minerals rich in water, aluminum, and silica, with sheetlike crystal structures. *Clay* also refers to any rock or mineral particle less than 0.00016 inch in diameter.

cleavage. Planes along which crystals or rocks break apart.

climate. The characteristic weather of a region, such as temperature and precipitation, averaged over some significant interval of time.

climatology. The scientific study of Earth's climate.

cobble. A rock fragment, usually worn and/or rounded, with a diameter between 2.5 and 10 inches.

conglomerate. A sedimentary rock consisting of pebbles, cobbles, or boulders cemented within a sandy matrix.

contact. The surface between rocks of two types or ages.

convection cell. The circulating motion of a fluid, generally driven by temperature differences; for example, convection cells develop in a pot of boiling water because the water at the bottom of the pot is hotter than that at the surface.

core-stone. A roughly spherical core of sound rock within partly disintegrated parent rock.

cosmogenic exposure dating. A geologic dating method that determines the amount of time a particular deposit, for example, a glacial erratic or rockfall boulder, has been resting at Earth's surface and exposed to cosmic rays.

crater. A steep-sided, circular depression commonly produced by an explosion.

crust. The outermost shell of Earth, consisting largely of the rock types exposed at the surface.

crystal. A many-faced solid, bound by smooth planar surfaces, that has an orderly internal arrangement of atoms.

debris flow. A typically fast-moving, liquefied flow of water-saturated, unconsolidated surface debris moving down a hillslope.

delta. A mass of sediment deposited by a stream into a standing body of water, such as an ocean or lake.

differential weathering. Differences in the degree of weathering a rock surface experiences, producing a surface that has variable topography and/or color. In Yosemite, it is common in granite that contains weathering-resistant minerals or enclaves.

dike. A tabular, intrusive igneous body of rock that cuts across the rock it intruded.

diorite. A coarse-grained intrusive igneous rock with more dark minerals than both granite and granodiorite.

dip. The inclination from horizontal of any planar surface (such as a layer) within rocks, as measured in the steepest direction; for example, the direction a marble would roll down the surface.

discharge. A measurement of flow in terms of volume per unit time. For a stream or river, it is the volume of water passing by a given spot in a second (cubic feet per second, or cfs).

divide. A topographic ridge separating water that flows into different drainage basins.

drainage basin. The area surrounded by drainage divides that steers tributary streams and rivers into a single main channel.

earthquake. Vibrations within Earth's crust produced by a sudden release of accumulated stress.

El Capitan Granite. A large pluton that makes up much of El Capitan and surrounding areas. It is characterized by biotite, a small amount of hornblende, and large (on the scale of an inch) irregular crystals of orthoclase.

electron. A subatomic particle with a negative charge that orbits the nucleus of an atom.

element. A substance composed of just one kind of matter that cannot be separated into different substances by chemical means.

enclave. A rounded blob of diorite or other rock, typically fist sized and dark, in another plutonic rock such as granite.

epidote. A mineral, commonly green, formed mostly by low-grade metamorphism.

erosion. The wearing away of soil and rock by weathering, mass wasting (landslides, rockfalls), and the action of streams, glaciers, waves, wind, and underground water.

erosion surface. A land surface, generally level or nearly level, shaped and subdued by erosion.

erratic. A boulder transported and deposited by a glacier.

exfoliation. The shedding of relatively thin layers of surface rock by the formation of joints that parallel the surface of the rock.

exfoliation joint. A crack that opens up at shallow depths below and roughly parallel to the surface of a rock slope.

extrusive rock. Rock formed from lava that is erupted onto the ground surface.

fault. A fracture along which blocks of Earth's crust have slipped past each other.

fault zone. A narrow zone of parallel or branching faults along which two blocks of crust slip past each other.

feldspar. A group of common, light-colored rock-forming minerals composed principally of silica, aluminum, and oxygen, plus one or more of the elements calcium, sodium, and potassium.

floodplain. A low-lying area adjacent to a river that is subject to flooding.

foliation. Mineralogical or textural banding in rocks that forms primarily during metamorphism in which the rocks stay solid.

formation. A rock body of considerable geographic extent with consistent characteristics that permit it to be recognized, mapped, and usually named.

fracture. Any break in rocks caused by natural mechanical failure under stress, including cracks, joints, and faults.

gaging station. Equipment used to measure height and other parameters of river flow.

garnet. A hard, dense, cleavage-free silicate mineral, typically with abundant aluminum, calcium, magnesium, and iron. It occurs in shades of red, brown, and green.

glacial erratic. See **erratic**

glacial polish. See **polish**

glacial till. See **till**

glaciation. A period during which a particular landscape was covered with glacial ice.

glacier. A body of ice that flows over land.

glass. A substance that forms when magma cools so rapidly that its atoms do not form a crystalline structure but rather are arranged randomly. It is not a mineral.

gneiss. Strongly metamorphosed rock characterized by alternating, irregular bands of coarse mineral grains and finer, flaky mica minerals.

gradient. For a river, the slope or incline of its bed.

granite. A light-colored, coarse-grained intrusive igneous rock consisting mostly of visible crystals of quartz and feldspar.

granodiorite. A coarse-grained intrusive igneous rock with more dark minerals than granite but less than diorite.

granodiorite of Kuna Crest. The outermost unit of the Tuolumne Intrusive Suite. It is composed of relatively dark granodiorite that can have well-aligned minerals.

groundmass. In an igneous rock, the fine-grained crystals or glass surrounding larger crystals.

grus. The granular product formed by the weathering of coarse-grained igneous rock; disintegrated granite.

Half Dome Granodiorite. The middle unit of the Tuolumne Intrusive Suite, composed of coarse-grained granodiorite with prominent black hornblende crystals nearly 1 inch long. It underlies the area between Glacier Point and Tenaya Lake.

hanging valley. A tributary valley with a floor that is distinctly higher than the main valley it joins.

hanging waterfall. A waterfall formed where a hanging valley joins a main valley.

hornblende. A complex rock-forming mineral in the amphibole group. It is commonly black with prismatic crystals.

hydrology. The scientific study of the properties, circulation, and distribution of water on and under Earth's surface and in the atmosphere.

Ice Age. The period between 2 million and 12,000 years ago during which large sheets of ice inundated parts of continents beyond their polar regions. Technically, it is known as the Pleistocene Epoch.

icefall. A feature similar to a waterfall but composed of ice. It forms where a glacier flows over a prominent bedrock step.

ice field. A very large, thick sheet of glacial ice generally flowing outward in all directions.

ice stream. A current of fast-flowing ice moving down a mountain valley or within a large ice sheet.

igneous rock. Rock formed by the crystallization of magma.

interglacial period. A warmer and possibly drier interval of time separating two glacial periods.

intrusive rock. Rock formed from magma that intrudes rock beneath the surface and then cools.

island arc. A series of volcanic islands, such as the Aleutian Islands, arranged in an arc above a subduction zone.

isotope. A species of an element defined by the number of neutrons in its nucleus. (Adjective: *isotopic*.)

isotopic dating. Determining the age of a geological sample by determining the ratios of certain isotopes it contains.

joint. A planar or near-planar fracture in bedrock along which no displacement has occurred. They often occur in parallel sets.

knickpoint. Any interruption or break in the relatively smooth profile of a streambed or riverbed, particularly a point of abrupt change (see **hanging valley**).

lateral moraine. An accumulation of glacial till along the lateral margins of a glacier. The accumulations remain as ridges or embankments after the glacier has retreated.

lava. Magma, or its solidified product, extruded onto Earth's surface.

limestone. A sedimentary rock composed largely of the mineral calcite.

limnology. The scientific study of freshwater bodies, especially lakes and ponds.

lode. A vein of ore.

luster. The manner in which light is reflected from a mineral surface; for example, glassy, greasy, or metallic.

magma. Molten rock.

magnitude. For earthquakes it is a measure of strain energy released during an event. Magnitude is measured on a logarithmic scale, with each increase in unit of magnitude corresponding to a ten-fold increase in amplitude of ground shaking and a thirty-fold increase in the energy released.

mantle. The part of Earth between the core and crust. Its upper portion is largely composed of the minerals olivine and pyroxene.

massive. Said of a rock that lacks internal structure or layers and has a homogenous composition.

matrix. The fine-grained rock or mineral particles filling the spaces between the coarser constituents of a sedimentary rock.

megacryst. An exceptionally large crystal in an igneous rock (typically more than 1 inch), such as orthoclase crystals in the Cathedral Peak Granite.

meltwater. Water derived from the melting of snow and glacial ice.

metamorphic rock. A rock that has been changed by heat, pressure, and stress so that it is distinct from its parent rock.

metamorphism. A change in the character of a rock, typically by recrystallization, as a result of heat and pressure.

mica. A group of silicate minerals with cleavage that causes them to form flakes or sheets.

mineral. A homogeneous, naturally occurring, inorganic solid substance with a specific chemical composition and specific physical properties.

moraine. A distinct ridge or series of ridges and mounds representing a concentrated accumulation of till that a glacier deposited while at a stable position. See also **lateral moraine, recessional moraine, terminal moraine**

mudstone. A fine-grained sedimentary rock composed of silt and clay. It is coarser grained and more massive than shale.

neutron. A subatomic particle with no electric charge in the nucleus of an atom.

North American Plate. The tectonic plate that includes most of North America. This plate's borders are roughly the West Coast of North America, the Mid-Atlantic Ridge along the center of the Atlantic Ocean, and a line running through Hispaniola, Cuba, and central Mexico. The northern, Arctic boundary is not well known.

nucleus. A cluster of protons and neutrons forming the central core of an atom.

orthoclase. A silicate mineral in the feldspar group that is rich in potassium and aluminum and common in igneous rocks.

oxidation. A chemical combination with oxygen. The oxidation of iron produces a red color in rocks.

Pacific Plate. The tectonic plate that underlies the Pacific Ocean and is west of the North American Plate.

paleo-. Pertaining to the geologic past.

pebbles. Small stones, usually worn to rounded, between 0.17 and 2.5 inches in diameter.

pegmatite. An abnormally coarse-grained igneous rock, typically in the form of a dike and commonly rich in silica.

placer. A mineral deposit formed at the surface by the mechanical concentration of mineral particles from weathered debris. They often form in streambeds.

plagioclase. A silicate mineral in the feldspar group that is rich in calcium and sodium. It is one of the most common minerals in igneous rocks.

plate. See **tectonic plate**

plate tectonics. The theory that Earth's surface is composed of tectonic plates that move independently of one another and interact at their boundaries, causing rock deformation, earthquakes, volcanic activity, and other phenomena.

plucking. An erosive process whereby sizable blocks of rock are loosened, picked up, and carried away by glaciers.

pluton. A body of igneous rock within Earth that cooled as one unit and has a relatively uniform composition.

polish. A smooth, shiny surface on rock formed by glacial abrasion.

porphyry. An igneous rock with some crystals that are much larger than others.

proton. A subatomic particle with a positive charge in the nucleus of an atom.

pumice. A highly cellular (bubble-filled) volcanic glass, rich in silica, that is often light enough to float on water.

pyrite. A brassy mineral composed of iron and sulfur. It forms cubes and is called "fool's gold" because of its color.

pyroclastic deposit. Broken rock material deposited during a volcanic explosion.

quartz. A common rock-forming mineral composed of silicon and oxygen. It can be clear, light gray, or brownish with a glassy or greasy luster.

quartzite. Principally, a metamorphic rock formed from quartz-rich sandstone.

radiocarbon dating. A method using a radioactive isotope of carbon (carbon-14) to determine the amount of time that has elapsed since a living organism died.

recessional moraine. An accumulation of till (a ridge or embankment) that a receding glacier deposited along its margins during a pause in its retreat.

rhyolite. An extrusive igneous rock, commonly light colored or reddish, that is relatively rich in silicon and poor in iron, magnesium, and calcium. Compositionally, it is the extrusive version of granite.

rift. A part of Earth's crust that has been stretched, such as the Basin and Range province of western North America.

rock avalanche. A rockfall or rockslide characterized by an especially large size, rapid emplacement velocity, excessive travel distance, and the distinctive hummocky terrain it creates.

rock debris. Broken-up or decomposing rock at the surface.

rockfall. Newly detached bedrock falling rapidly from a cliff or other steep surface.

rockslide. Newly detached bedrock sliding down a rock surface.

salinity. A measure of the concentration of salt in a solution, such as seawater.

sandstone. A sedimentary rock composed primarily of sand-sized particles (0.0025 to 0.08 inch in diameter) of rock or mineral.

schist. A metamorphic rock characterized by well-defined, thin foliation usually involving the orientation of flakes of mica.

sediment. Unconsolidated particulate matter deposited by some agent of transport, for example, wind or water.

sedimentary rock. Consolidated sediment characterized by layering and usually held together by natural cement.

seismic imaging. Determining the configuration of rock below the surface using seismic waves.

seismic waves. Elastic waves, such as sound, produced in Earth by events such as earthquakes, explosions, and rockfalls.

seismometer. A device for recording seismic waves.

serpentine. A green, fibrous silicate mineral that is rich in magnesium and produced by the metamorphism of olivine.

shale. A fine-grained, layered sedimentary rock composed of mud and sand.

silicate. A compound (typically a mineral) built in part of silicon and oxygen atoms.

silt. Fine, dust-sized (finer than sand, coarser than clay) rock and mineral matter between 0.00016 and 0.0025 inch in diameter.

slate. Slightly metamorphosed shale that is harder than shale and has pronounced cleavage.

sphene. A silicate mineral rich in calcium and titanium. It typically forms tiny, honey-colored crystals in granite.

spillway. A channel, such as the outlet of a lake, that carries excess water over or around an obstruction or out of a body of water.

stage. In hydrology, the average height of a river.

strandline. A shoreline carved by wind-driven waves when a standing body of water, such as a lake, sat at a particular level for a long time.

striations. Linear, parallel scratches on a polished rock surface, or linear, parallel marks on a mineral surface (particularly plagioclase and pyrite) caused by the mineral's internal structure.

strike. The compass direction a horizontal line points within a planar feature, such as a rock layer.

subduction. A process in which one tectonic plate descends beneath another, producing magma.

subduction zone. A tectonic setting in which one tectonic plate descends beneath another tectonic plate.

supercontinent. An aggregation of most of Earth's continents in one continental mass.

Tahoe glaciation. The local name for the glaciation that occurred in the Sierra Nevada roughly 140,000 to 80,000 years ago.

talus. An accumulation of coarse, angular blocks of rock at the base of a cliff or other steep bedrock slope. The product of rockfalls and rockslides.

tectonic. Said of forces involved in the deformation of Earth's crust.

tectonic deformation. Faults, folds, tilting, and uplift created in rocks by stresses in Earth's crust.

tectonic plate. In a planetary sense, one of eight large drifting plates composing Earth's solid outer part. Plates are roughly 60 miles thick.

terminal moraine. An accumulation of glacial till dumped at the end of a glacier that remains as a ridge or embankment after the glacier has retreated. It marks the farthest reach the glacier achieved.

terrane. A large region underlain by rocks with similar characteristics and a common history.

texture. The physical characteristics of a rock, such as the size, alignment, and layering of its minerals.

till. Jumbled rock debris (fine and coarse) deposited directly from a glacier.

Tioga glaciation. The local name for the most recent large glaciation in the Sierra Nevada that occurred between 26,000 and 18,000 years ago during a prolonged cold period.

tributary. A small stream that flows into a larger stream.

tufa. A rock composed mainly of calcium carbonate or silica precipitated from water and found at spring sites or along the strandlines of saline lakes.

Tuolumne Intrusive Suite. A large, nested family of granite and granodiorite plutons of Cretaceous age that makes up much of the eastern half of Yosemite National Park. The suite comprises, from oldest to youngest, the granodiorite of Kuna Crest, Half Dome Granodiorite, and Cathedral Peak Granodiorite.

vein. A sheetlike deposit of a mineral or minerals within a fracture in rock.

vent (volcanic). A nearly vertical, roughly cylindrical opening through which volcanic material is extruded.

viscous. The state of a fluid with a cohesive, sticky consistency. A viscous fluid flows sluggishly, whereas a nonviscous fluid flows easily.

volcanic ash. Unconsolidated, explosively fragmented volcanic material with a particle diameter of less than 0.125 inch.

volcanic glass. See **glass**

volcanic rock. Glassy to finely crystallized lava, or fragmented lava debris, erupted from a volcanic vent.

weathering. The in-place physical (breaking up) and chemical (altering the chemical makeup) degradation of rock, usually accomplished by water.

xenolith. A rock fragment caught up in magma that is unrelated to it.

Sources of More Information

GENERAL READING

Bateman, P. C., and C. Wahrhaftig. 1966. Geology of the Sierra Nevada. In *California Division of Mines Bulletin* 190, ed. E. H. Bailey, 107–72.

Hill, M. 2006. *Geology of the Sierra Nevada*. California Natural History Guides 80. Berkeley and Los Angeles: University of California Press.

———. 2007. *Geological Ramblings in Yosemite*. Berkeley, Calif.: Heyday Books.

Matthes, F. E. 1930. *Geologic History of the Yosemite Valley*. U.S. Geological Survey Professional Paper 160.

———. 1950. *The Incomparable Valley: A Geologic Interpretation of the Yosemite*. Ed. F. Fryxell. Berkeley: University of California Press.

Muir, Jo 1873. On actual glacier in California. *American Journal of Science and Arts*. 5:69–71.

Muir, J. 1912. *The Yosemite*. New York: The Century Co.

Sharp, R. P., and A. F. Glazner. 1997. *Geology Underfoot in Death Valley and Owens Valley*. Missoula, Mont.: Mountain Press Publishing Co.

INTRODUCTION

Baldridge, W. S. 2004. *Geology of the American Southwest: A Journey Through Two Billion Years of Plate Tectonic History*. Cambridge, UK: Cambridge University Press.

Blakey, R., and W. Ranney. 2008. *Ancient Landscapes of the Colorado Plateau*. Grand Canyon, Ariz.: Grand Canyon Association.

Farquhar, F. P., ed. 2003. *Up and Down California in 1860–1864: The Journal of William H. Brewer*. Berkeley: University of California Press.

Ghiglieri, M. P., and C. R. Farabee Jr. 2007. *Off the Wall: Death in Yosemite*. Flagstaff, Ariz.: Puma Press.

Gutenberg, B., J. P. Buwalda, and R. P. Sharp. 1956. Seismic explorations on the floor of Yosemite Valley, California. *Geological Society of America Bulletin* 67:1051–78.

Guyton, B. 1998. *Glaciers of California: Modern Glaciers, Ice Age Glaciers, the Origin of Yosemite Valley, and a Glacier Tour in the Sierra Nevada*. California Natural History Guides 59. Berkeley: University of California Press.

Huber, N. K. 1989. *Geologic Story of Yosemite National Park*. Yosemite National Park: Yosemite Association.

Huber, N. K. 2007. A Tale of Two Valleys. In *Geological Ramblings in Yosemite*, 79–85. Berkeley, Calif.: Heyday Books.

Jones, C. H., G. L. Farmer, and J. Unruh. 2004. Tectonics of Pliocene removal of lithosphere of the Sierra Nevada, California. *Geological Society of America Bulletin* 116:1408–22.

MacGregor, K. R., R. S. Anderson, S. P. Anderson, and E. D. Waddington. 2000. Numerical simulations of glacial-valley longitudinal profile evolution. *Geology* 28:1031–34.

Mount, J. F. 1995. *California Rivers and Streams: The Conflict Between Fluvial Process and Land Use.* Berkeley: University of California Press.

Mulch, A., S. A. Graham, and C. P. Chamberlain. 2006. Hydrogen isotopes in Eocene river gravels and paleoelevation of the Sierra Nevada. *Science* 313:87–89.

Ruddiman, W. 2001. *Earth's Climate: Past and Future.* New York: W. H. Freeman.

Sharp, R. P. 1988. *Living Ice: Understanding Glaciers and Glaciation.* Cambridge, UK: Cambridge University Press.

Wakabayashi, J., and T. L. Sawyer. 2001. Stream incision, tectonics, uplift, and evolution of topography of the Sierra Nevada, California. *Journal of Geology* 109:539–62.

1. Bedrock

Bateman, P. C., and B. W. Chappell. 1979. Crystallization, fractionation, and solidification of the Tuolumne Intrusive Series, Yosemite National Park, California. *Geological Society of America Bulletin* 90:465–82.

Bateman, P. C. 1992. *Plutonism in the Central Part of the Sierra Nevada Batholith, California.* U.S. Geological Survey Professional Paper 1483.

Coleman, D. S., W. Gray, and A. F. Glazner. 2004. Rethinking the emplacement and evolution of zoned plutons: Geochronologic evidence for incremental assembly of the Tuolumne Intrusive Suite, California. *Geology* 32:433–36.

Gray, W., A. F. Glazner, D. S. Coleman, and J. M. Bartley. 2008. Long-term geochemical variability of the Late Cretaceous Tuolumne Intrusive Suite, central Sierra Nevada, California. *Geological Society, London, Special Publication* 304:183–201.

Ratajeski, K., A. F. Glazner, and B. V. Miller. 2001. Geology and geochemistry of mafic to felsic plutonic rocks associated with the Cretaceous intrusive suite of Yosemite Valley, California. *Geological Society of America Bulletin* 113:1486–1502.

2. Geology of Climbing

Barnes, G., C. McNamara, S. Roper, and T. Snyder. 2003. *Yosemite Valley Free Climbs: Supertopos.* Mill Valley, Calif.: SuperTopo LLC.

Garlick, S. 2009. *Flakes, Jugs, and Splitters: A Rock Climber's Guide to Geology.* How to Climb Series. Guilford, Conn.: Falcon Guides.

Huber, N. K. 2007. A Tale of Two Valleys. In *Geological Ramblings in Yosemite*, 79–85. Berkeley, Calif.: Heyday Books.

Rowell, G. A., ed. 1974. *The Vertical World of Yosemite: A Collection of Writings and Photographs on Rock Climbing in Yosemite.* Berkeley, Calif.: Wilderness Press.

3. Yosemite Falls

Huber, N. K. 2007. Yosemite Falls—A New Perspective. In *Geological Ramblings in Yosemite*, 95–102. Berkeley, Calif.: Heyday Books.

Matthes, F. E. 1930. *Geologic History of the Yosemite Valley*. U.S. Geological Survey Professional Paper 160.

Osborne, M. 2009. *Granite, Water, and Light: The Waterfalls of Yosemite Valley*. Berkeley, Calif.: Heyday Books.

4. Vernal and Nevada Falls

(see Matthes and Osborne references for **3. Yosemite Falls**)

Sharp, R. P. 1988. *Living Ice: Understanding Glaciers and Glaciation*. Cambridge UK: Cambridge University Press.

5. Happy Isles

Snyder, J. B. 1996. The Ground Shook and the Sky Fell. *Yosemite* 58 (Fall):2–9.

Wieczorek, G. F., and S. Jäger. 1996. Triggering mechanisms and depositional rates of post-glacial slope-movement processes in the Yosemite Valley, California. *Geomorphology* 15 (1):17–31.

Wieczorek, G. F., J. B. Snyder, R. B. Waitt, M. M. Morrissey, R. A. Uhrhammer, E. L. Harp, R. D. Norris, M. I. Bursik, and L. G. Finewood. 2000. Unusual July 10, 1996, rock fall at Happy Isles, Yosemite National Park, California. *Geological Society of America Bulletin* 112:75–85.

Wieczorek, G. F., and J. B. Snyder. 2004. *Historical Rock Falls in Yosemite National Park, California*. U.S. Geological Survey Open-File Report 03–491.

6. Cookie Cliff Rockslide

(see Wieczorek and Snyder, 2004 for **5. Happy Isles**)

7. Rock Avalanches

Lee, J., J. Spencer, and L. Owen. 2001. Holocene slip rates along the Owens Valley fault, California: Implications for the recent evolution of the Eastern California Shear Zone. *Geology* 29:819–22.

Stock, G. M., and R. A. Uhrhammer. 2010. Catastrophic rock avalanche 3,600 years B.P. from El Capitan, Yosemite Valley, California. *Earth Surface Processes and Landforms* (in press).

Wieczorek, G. F., M. M. Morrisey, G. Iovine, and J. Godt. 1999. *Rock-Fall Potential in the Yosemite Valley, California*. U.S. Geological Survey Open-File Report 99–578.

Wieczorek, G. F. 2002. Catastrophic Rockfalls and Rockslides in the Sierra Nevada, USA. In *Catastrophic Landslides: Effects, Occurrence, and Mechanisms*, ed. S. G. Evans and J. V. DeGraff, 165–90. Boulder, Colo.: Geological Society of America.

8. Flood of 1997

Mount, J. F. 1995. *California Rivers and Streams: The Conflict Between Fluvial Processes and Land Use*. Berkeley: University of California Press.

Rylands, K. 1997. The flood of the century. *Yosemite* 59 (Spring):1–6.

9. El Capitan Moraine

Huber, N. K. 2007. A History of the El Capitan Moraine. In *Geological Ramblings in Yosemite*, 103–10. Berkeley, Calif.: Heyday Books.

Matthes, F. 1962. El Capitan Moraine and Ancient Lake Yosemite. In *François Matthes and the Marks of Time: Yosemite and the High Sierra*, ed. F. Fryxell, 55–62. San Francisco, Calif.: Sierra Club.

10. Taft Point

Ericson, K., P. Mignon, and M. Olymo. 2005. Fractures and drainage in the granite mountainous area: a study from the Sierra Nevada, USA. *Geomorphology* 64:97–116.

Huber, N. K. 1989. *Geologic Story of Yosemite National Park*. Yosemite National Park: Yosemite Association.

Matthes, F. E. 1930. *Geologic History of the Yosemite Valley*. U.S. Geological Survey Professional Paper 160.

11. Sentinel Dome

(see references for **10. Taft Point**)

Gilbert, G. K. 1904. Domes and dome structure of the High Sierra. *Geological Society of American Bulletin* 15:29–36.

12. Exfoliation

Huber, N. K. 1989. *Geologic Story of Yosemite National Park*. Yosemite National Park: Yosemite Association.

Jahns, R. H. 1943. Sheet structure in granites: its origin and use as a measure of glacial erosion in New England. *Journal of Geology* 51:71–98.

Martel, S. J. 2006. Effect of topographic curvature on near-surface stresses and application to sheeting joints. *Geophysical Research Letters* 33 (1), L01308.

Matthes, F. E. 1930. *Geologic History of the Yosemite Valley*. U.S. Geological Survey Professional Paper 160.

13. Pothole Dome

Balogh, R. S. 1987. Pothole Dome: where water flowed uphill, Tuolumne County. *California Geology* 40 (July):154–57.

Gilbert, G. K. 1906. Moulin work under glaciers. *Geological Society of America Bulletin* 17:317–20.

Sharp, R. P. 1988. *Living Ice: Understanding Glaciers and Glaciation*. Cambridge, UK: Cambridge University Press.

14. Olmsted Point

Guyton, B. 1998. *Glaciers of California: Modern Glaciers, Ice Age Glaciers, the Origin of Yosemite Valley, and a Glacier Tour in the Sierra Nevada*. California Natural History Guides 59. Berkeley: University of California Press.

Huber, N. K. 1989. *The Geologic Story of Yosemite National Park*. Yosemite National Park: Yosemite Association.

Matthes, F. E. 1930. *Geologic History of the Yosemite Valley*. U.S. Geological Survey Professional Paper 160.

Sharp, R. P. 1988. *Living Ice: Understanding Glaciers and Glaciation*. Cambridge, UK: Cambridge University Press.

15. Tenaya Lake

Matthes, F. 1962. The Scenery about Tenaya Lake. In *François Matthes and the Marks of Time: Yosemite and the High Sierra*, ed. F. Fryxell, 129–36. San Francisco, Calif.: Sierra Club.

Stine, S. 1994. Extreme and persistent drought in California and Patagonia during mediaeval time. *Nature* 369:546–49.

16. Soda Springs

Barnes, I., R. W. Kistler, R. H. Mariner, and T. S. Presser. 1981. *Geochemical Evidence on the Nature of the Basement Rocks of the Sierra Nevada, California*. U.S. Geological Survey Water-Supply Paper 2181.

O'Neill, E. S. 1984. *Meadow in the Sky: A History of Yosemite's Tuolumne Meadows Region*. Groveland, Calif.: Albicaulis Press.

17. May Lake

Lahren, M. M., R. A. Schweickert, J. M. Mattinson, and J. D. Walker. 1990. Evidence of uppermost Proterozoic to Lower Cambrian miogeoclinal rocks and the Mojave-Snow Lake fault: Snow Lake pendant, central Sierra Nevada, California. *Tectonics* 9:1585–1608.

Schweickert, R. A., and M. M. Lahren. 1991. Age and Tectonic Significance of Metamorphic Rocks along the Axis of the Sierra Nevada Batholith: A Critical Reappraisal. In *Paleozoic Paleogeography of the Western United States II*, ed. J. D. Cooper and C. H. Stevens, 653–76. Los Angeles, Calif.: Society of Economic Paleontologists and Mineralogists, Pacific Section.

18. Little Devils Postpile

Calk, L. C., and C. W. Naeser. 1973. The thermal effect of a basalt intrusion on fission tracks in quartz monzonite. *Journal of Geology* 81:189–98.

19. Metamorphic Rocks of Western Approaches

Bateman, P. C., and C. Wahrhaftig. 1966. Geology of the Sierra Nevada. In *California Division of Mines Bulletin* 190, ed. E. H. Bailey, 105–72.

Harp, E. L., M. E. Reid, J. W. Godt, J. V. DeGraff, and A. J. Gallegos. 2008. Ferguson rock slide buries California State Highway near Yosemite National Park. *Landslides* 5:331–37.

Snow, C. A., and H. Scherer. 2006. Terranes of the western Sierra Nevada foothills metamorphic belt, California: A critical review. *International Geology Review* 48:46–62.

20. Stanislaus Table Mountain

Bateman, P. C., and C. Wahrhaftig. 1966. Geology of the Sierra Nevada. In *California Division of Mines Bulletin* 190, ed. E. H. Bailey, 105–72.

Garside, L. J., C. D. Henry, J. E. Faulds, N. H. Hinz. 2005. The Upper reaches of the Sierra Nevada auriferous gold channels. In *Window to the World: Geological Society of Nevada Symposium Proceedings, May 14–18, 2005,* ed. N. H. Rhoden et al.

Pluhar, C. J., A. L. Deino, N. M. King, C. Busby, B. P. Hausback, T. Wright, and C. Fischer. 2009. Lithostratigraphy, magnetostratigraphy, and radiometric dating of the Stanislaus Group, CA, and age of the Little Walker Caldera. *International Geology Review* 51:873–99.

Rhodes, D. D. 1986–88. Table Mountain of Calaveras and Tuolumne Counties, California. In *Cordilleran Section of the Geological Society of America*, Vol. 1 of *The Decade of North American Geology: Centennial Field Guide*, ed. M. L. Hill, 269–72. Boulder, Colo.: Geological Society of America.

21. Eocene Erosion (Rim of the World)

Bateman, P. C., and C. Wahrhaftig. 1966. Geology of the Sierra Nevada. In *California Division of Mines Bulletin* 190, ed. E. H. Bailey, 105–72.

Cecil, M. R., M. N. Ducea, P. W. Reiners, and C. G. Chase. 2006. Cenozoic exhumation of the northern Sierra Nevada, California, from (U-Th)/He thermochronology. *Geological Society of America Bulletin* 118:1481–88.

Clark, M. K., G. Maheo, J. Saleeby, and K. A. Farley. 2005. The non-equilibrium landscape of the southern Sierra Nevada, California. *GSA Today* 15 (9):4–10.

Stock, G. M., R. S. Anderson, and R. C. Finkel. 2005. Rates of erosion and topographic evolution of the Sierra Nevada, California, inferred from cosmogenic ^{26}Al and ^{10}Be concentrations. *Earth Surface Processes and Landforms* 30:985–1006.

22. Mono Lake

Gaines, D. 1989. *Mono Lake Guidebook*. Lee Vining, Calif.: Kutsavi Books.

Reheis, M. C., S. Stine, and A. M. Sarna-Wojcicki. 2002. Drainage reversals in Mono Basin during the late Pliocene and Pleistocene. *Geological Society of America Bulletin* 114:991–1006.

Stine, S. 1990. Late Holocene fluctuations of Mono Lake, eastern California. *Palaeoclimatology, Palaeogeography, Palaeoecology* 78:333–81.

23. Black Point

Bursik, M., and K. E. Sieh. 1989. Range front faulting and volcanism in the Mono basin, eastern California. *Journal of Geophysical Research* 94:15,587–609.

Christensen, M. N., and C. M. Gilbert. 1964. Basaltic cone suggests constructional origin of some guyots. *Science* 143:240–42.

24. Lee Vining moraines

Dühnforth, M., R. S. Anderson, D. Ward, and G. M. Stock. 2010. Bedrock fracture control of glacial erosion processes and rates. *Geology.* 38:423–26.

Guyton, B. 1998. *Glaciers of California: Modern Glaciers, Ice Age Glaciers, the Origin of Yosemite Valley, and a Glacier Tour in the Sierra Nevada.* California Natural History Guides 59. Berkeley: University of California Press.

Huber, N. K. 2007. Exotic Boulders at Tioga Pass. In *Geological Ramblings in Yosemite*, 111–14. Berkeley, Calif.: Heyday Books.

Putnam, W. C. 1950. Moraine and shoreline relationships at Mono Lake, California. *Bulletin of the Geological Society of America* 61:115–22.

25. Bennettville and Dana Village

Hubbard, D. 1958. *Ghost Mines of Yosemite.* Fredericksburg, Tex.: Awani Press.

Patera, A. H. 2003. *Bennettville, and the Tioga Mining District.* Western Places no. 25. Lake Grove, Oreg.: Western Places.

Trent, D. D. 1986–88. Multiple Deformation in the Bennettville Area of the Saddlebag Lake Pendant, Central Sierra Nevada, California. In *Cordilleran Section of the Geological Society of America*, Vol. 1 of *The Decade of North American Geology: Centennial Field Guide*, ed. M. L. Hill, 273–76. Boulder, Colo.: Geological Society of America.

Geologic Maps

Alpha, T. R., C. Wahrhaftig, and N. K. Huber. 1987. Oblique map showing maximum extent of 20,000-year-old (Tioga) glaciers, Yosemite National Park, central Sierra Nevada, California. *Miscellaneous Investigations Series Map* I-1885, U.S. Geological Survey.

Calkins, F. C., J. A. Roller, and N. K. Huber. 1985. Bedrock geology map of Yosemite Valley, Yosemite National Park, California. *Miscellaneous Investigations Series Map* I-1639, U.S. Geological Survey.

Huber, N. K., P. C. Bateman, and C. Wahrhaftig. 1989. Geologic map of Yosemite National Park and vicinity. *Miscellaneous Investigations Series Map* I-1874, U.S. Geological Survey.

Index

Page numbers in *italics* indicate an illustration or information in a caption.

About the Authors

—Laura Flegg photo, copyright of Pioneer Productions

Allen F. Glazner is the Kenan Distinguished Professor of Geological Sciences at the University of North Carolina at Chapel Hill. A native Californian, he holds a PhD from the University of California at Los Angeles and has done research in the Sierra Nevada and Mojave Desert since his undergraduate days at Pomona College.

—Steven Bumgardner photo

Greg M. Stock was born and raised in the Sierra Nevada just north of Yosemite. His interest in the local geology led him to pursue a bachelor's degree in geology at Humboldt State University and a PhD in earth science at the University of California at Santa Cruz, where he studied the uplift and erosion of the Sierra Nevada. Greg is Yosemite National Park's first-ever park geologist. He lives in Yosemite Valley with his wife, Sarah, and daughter, Autumn.